The Basics of Spectroscopy

Tutorial Texts Series

The Basics of Spectroscopy

David W. Ball

Tutorial Texts in Optical Engineering
Volume TT49

Arthur R. Weeks, Jr., Series Editor
Invivo Research Inc. and University of Central Florida

SPIE PRESS
A Publication of SPIE—The International Society for Optical Engineering
Bellingham, Washington USA

Library of Congress Cataloging-in-Publication Data

Ball, David W. (David Warren), 1962-
    The basics of spectroscopy / David W. Ball
      p. cm. -- (Tutorial texts in optical engineering ; v. TT49)
    Includes bibliographical references (p.).
    ISBN 0-8194-4104-X (pbk. : alk. paper)
      1. Spectrum analysis. I. Title. II. Series.

    QC451 .B18 2001
    543'.0858–dc21

                                                    2001032208
                                                          CIP

Published by

SPIE—The International Society for Optical Engineering
P.O. Box 10
Bellingham, Washington 98227-0010 USA
Phone: 360/676-3290
Fax: 360/647-1445
Email: spie@spie.org
WWW: www.spie.org

Printed in the United States of America.

For my sons, who span the Millennium:
Stuart Ryan (b. 9/99)
and
Alex Casimir ("Casey", b. 2/01)

Introduction to the Series

The Tutorial Texts series was initiated in 1989 as a way to make the material presented in SPIE short courses available to those who couldn't attend and to provide a reference book for those who could. Typically, short course notes are developed with the thought in mind that supporting material will be presented verbally to complement the notes, which are generally written in summary form, highlight key technical topics, and are not intended as stand-alone documents. Additionally, the figures, tables, and other graphically formatted information included with the notes require further explanation given in the instructor's lecture. As stand-alone documents, short course notes do not generally serve the student or reader well.

Many of the Tutorial Texts have thus started as short course notes subsequently expanded into books. The goal of the series is to provide readers with books that cover focused technical interest areas in a tutorial fashion. What separates the books in this series from other technical monographs and textbooks is the way in which the material is presented. Keeping in mind the tutorial nature of the series, many of the topics presented in these texts are followed by detailed examples that further explain the concepts presented. Many pictures and illustrations are included with each text, and where appropriate tabular reference data are also included.

To date, the texts published in this series have encompassed a wide range of topics, from geometrical optics to optical detectors to image processing. Each proposal is evaluated to determine the relevance of the proposed topic. This initial reviewing process has been very helpful to authors in identifying, early in the writing process, the need for additional material or other changes in approach that serve to strengthen the text. Once a manuscript is completed, it is peer reviewed to ensure that chapters communicate accurately the essential ingredients of the processes and technologies under discussion.

During the past nine years, my predecessor, Donald C. O'Shea, has done an excellent job in building the Tutorial Texts series, which now numbers nearly forty books. It has expanded to include not only texts developed by short course instructors but also those written by other topic experts. It is my goal to maintain the style and quality of books in the series, and to further expand the topic areas to include emerging as well as mature subjects in optics, photonics, and imaging.

Arthur R. Weeks, Jr.
Invivo Research Inc. and University of Central Florida

CONTENTS

PREFACE

This book is largely based on a series of essays published as "The Baseline" column in the trade periodical *Spectroscopy*. I am indebted to the editor of *Spectroscopy*, Mike MacRae, and its editorial board for granting permission to reprint and/or adapt the columns for the purpose of this book.

The discussions that led to the inception of "The Baseline" were based on a growing understanding by *Spectroscopy*'s editorial staff that readers of the magazine were suffering from a lack of basic, tutorial-style information about spectroscopy, its theories, its applications, and its techniques. Most of the readership did have some sort of technical education, but it was (a) varied, and (b) in the past. Many readers felt that they would benefit from short, simple articles that covered "how-and-why" topics in spectroscopy. And so, "The Baseline" was born.

Having participated in some of the discussions myself, I eagerly volunteered to pen the columns. Writing such a column appealed to me in several ways. First, it appealed to the teacher in me. A new classroom, a new audience, a new way to spread the word spectroscopic! Second, I recognized the truism that you learn more when you write about it. In the past 6-plus years, I have learned more from writing these columns and receiving feedback about them than I ever would from studying an instrument manual. Finally, I must confess to being a huge fan of Isaac Asimov. I learned a lot by reading (and rereading and rereading…) his essays on science et al., and I am ecstatic at the opportunity to emulate my science-writing hero. (At least in some respects.) To date, more than two dozen columns have appeared in print, most of them written by me. And to be honest, over time I wondered if there would ever be the opportunity to print a collection of the columns in book form—another emulation of my science-writing hero.

With the exception of pointing out a minor error here and there (and I hope they have all been corrected for this book!), the feedback I have received from the readers has been universally positive. Several people have been in touch regularly because of the column, and I've been contacted by old friends and colleagues who, after years of separation, see my name. It's been a great thing.

In December 1999, Eugene Arthurs, Executive Director of SPIE, contacted me with the proposal to reprint the columns, properly revised, in book form. It would become part of SPIE's Tutorial Text Series. It didn't take much review of some of the already published Tutorial Texts to realize that "The Baseline" and the Tutorial Text Series are an excellent match. You are holding the end product.

Thanks to Eugene Arthurs for his interest and support. Thanks also to Sherry Steward and Mike MacRae, the editors at *Spectroscopy*, and all the associate and assistant editors who have helped keep "The Baseline" column going. Bradley M. Stone (San Jose State University) and another anonymous reviewer read the manuscript, corrected several minor errors, and found many mistakes that were ultimately derived from the voice-recognition software that I used to regenerate some of the earlier columns that were longer available in electronic form. Timothy Lamkins at SPIE Press read the manuscript and pointed out a few inconsistencies that inevitably remained. Finally, Rick Hermann and Merry Schnell, also at SPIE Press, were my main contacts there and offered valuable advice.

The Basics of Spectroscopy is not a detailed, high-level mathematical, rigorous treatment of spectroscopy. Rather, it is an easy-reading, tutorialized treatment of some of the basic ideas of the field. (In fact, every chapter could be expanded into several books' worth of material that focused on that particular topic. A quick scan of any university library's shelves will confirm that.) The level of vernacular is *not* meant to sacrifice accuracy; rather, it is meant to improve comprehension, especially by readers who might not be graduate-level-trained scientists and engineers. The better that readers can grasp the basics of the topic, the better chance they have to understand the details of the topics—and those can be found in textbooks, technical articles, (sometimes) manuals, and so on. There are plenty of those in libraries and classrooms, if you really want to find them—some of them are listed as references at the ends of the chapters. *Basics* is a possible first step for those who want to know more about spectroscopy.

Because the book is based on a series of columns, there may be a rather unsystematic feel to the presentation of the material. While I have done my best to make for smooth transitions, the reader should keep in mind that this book is based on 1000-word essays on different topics. I have grouped similar topics together in a way that hopefully makes sense, and I've added some previously unpublished material to fill in any major gaps. Of course, not all the gaps are filled, but it is impossible to fill all of them with a book like this. Again, the reader is encouraged to consider higher-level sources, once this book whets one's appetite.

The book starts with an abbreviated history of light and spectroscopy, then discusses the interaction of light with matter. Spectrometer basics are introduced next, followed by a discussion of a spectrum itself. This is followed by quantitative and qualitative aspects of a spectrum, a brief (as it must be!) discussion of quantum mechanics, selection rules, and experimental factors. The book weaves basic topics of physics and physical chemistry, analytical chemistry, and optics into one volume.

I hope that, from the reader's perspective and in light of its intended scope, this book serves its purpose well.

David W. Ball
Cleveland State University
Cleveland, Ohio
April 2001

NOMENCLATURE

Vectors are denoted by **boldface**.
Quantum-mechanical operators are denoted with a caret ^ over them:

λ – wavelength
$\tilde{\nu}$ – wavenumber
ν – frequency
h – Planck's constant
$\hbar$ – Planck's constant divided by 2π
B – Einstein coefficient of stimulated absorption
R_H – Rydberg's constant
σ – Stefan-Boltzmann constant
k – Boltzmann constant
c – speed of light
Ψ – wavefunction
B – magnetic field
E – electric field
a_0 – radius of first Bohr radius
i – the square root of –1: $\sqrt{-1}$
P – power
T – transmittance or temperature
A – absorbance
I – intensity
I – quantum number for nuclei
S – quantum number for electrons
l – angular momentum quantum number
j – total electronic angular momentum quantum number
m_S – z-component of total spin angular momentum for electrons
m_I – z-component of total spin angular momentum for nuclei
ε – molar absorptivity
ε_0 – permittivity of free space
e – charge of electron
n – refractive index
α – absorption coefficient
γ – magnetogyric ratio
κ – attenuation factor
β – nuclear magneton
δ – optical path difference
μ – magnetic moment
p – momentum
M – transition moment
$\hat{H}$ – Hamiltonian operator

The Basics of
Spectroscopy

Chapter 1
A SHORT HISTORY

1.1 Introduction

Spectroscopy is the study of matter using electromagnetic radiation. While this definition is nominally correct, it is rather simple. On this basis, one could argue that everything we know about the universe comes from spectroscopy, since much of we have learned comes from what we see in the world around us. But simply looking at a picture or painting is not usually considered "spectroscopy," even though the action might involve studying a piece of matter in broad daylight.

While we will not attempt to develop a more detailed definition of spectroscopy in the remainder of this book, we will be examining various aspects of spectroscopy that make it a scientific tool. In order to set the stage better for the various topics that will be presented, we present a quick history of the development of topics relevant to spectroscopy. There are three major topics: matter, light, and the fusion of matter and light that was ultimately (and properly) labeled "spectroscopy."

1.2 Matter

Throughout most of history, matter was assumed to be *continuous*—that is, you could separate it into increasingly smaller pieces, and each piece could then be cut into smaller and smaller parts, ad infinitum. Common experience shows that to be the case, doesn't it? Furthermore, ancient philosophers (as thinkers were known at the time) divided matter into several fundamental substances that were subject to various mystical forces. The four fundamental substances, or *elements*—fire, air, water, and earth—had accompanying attributes—wet, dry, cold, and hot—that they imparted to matter, depending on the relative amounts in each object. Such a description of matter is attributed to the fifth-century B.C. philosopher Empedocles. Figure 1.1 shows the relationship between the four elements and their attributes. Plato and his pupil Aristotle (fifth to fourth century B.C.) supported these ideas and refined them (in part by introducing a fifth "heavenly" element, the ether). Because of Plato's and Aristotle's influence on the thinking of the time (and times since), the "four elements" idea of matter was the prevailing view for centuries in the Western world. (Three additional medical *principles*—sulfur, salt, and mercury—were added to the repertoire by the sixteenth-century physician Paracelsus.)

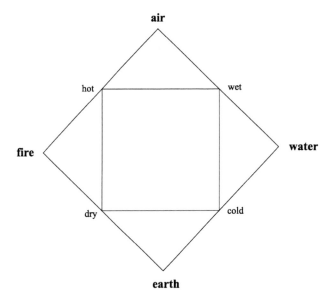

Figure 1.1 According to the four-elements description of matter, all matter was composed of four basic elements: earth, air, fire, and water. Different matter had different proportions of each. Each element also imparted certain attributes to the matter, like hot or cold or wet or dry. (Adapted from Ihde, *The Development of Modern Chemistry*, Dover Press.)

A competing description of matter was proposed at about the same time, however. In the fifth to fourth century B.C., Democritus proposed (based on ideas from his teacher, Leucippus) that matter was ultimately composed of tiny, solid particles named atoms. However, this idea never found favor because of Aristotle's influential support of other viewpoints. (Throughout history and even in modern times, influential thinkers use their influence to sway the direction of scientific thought.) Besides, common experience shows that matter is not made up of tiny particles—it is continuous!

It was not until the seventeenth century that the concept of matter began to change. This change was prompted by two interconnected events. First, what we now call the scientific method—a more formalized methodology for studying the natural universe[1] – was being promoted by people like Sir Francis Bacon and, from a more philosophic perspective, René Descartes. Eventually, a less haphazard and more systematic approach toward the study of matter began to percolate through the community of natural

---

[1] Since the details of the scientific method are available elsewhere, we will not present them here, and assume that the reader is familiar with its general ideas.

philosophers. Second, work by Robert Boyle in the mid-seventeenth century on the physical properties of gases revived the idea of matter as atoms as a model to explain gases' pressure-volume behavior. In fact, Boyle's work on gases can be thought of as the dividing line between old-style alchemy and the beginning of modern chemical investigations.

Based on a century of new work and ideas, in 1789, Antoine Lavoisier published *Traité élémentaire de chemie* ("Elements of Chemistry"). In it, the four-elements idea of the ancients is replaced by another definition of element: a substance that cannot be simplified further by chemical means. Not only did Lavoisier publish a table listing substances he recognized as elements (and some that we now do not recognize as elements, like lime and magnesia), but he also showed that water isn't an element by making it from hydrogen and oxygen. Ideas in science don't change overnight, but in time Lavoisier's views became prevalent, and the "four elements" concept of matter was eventually replaced.

With the results of almost two centuries of scientific-method-based inquiry in hand, in 1803, John Dalton began to enunciate his atomic theory of atoms. All matter is composed of tiny indivisible particles called atoms (a word borrowed from Democritus). All atoms of the same element are the same, while atoms of different elements are different, and atoms of different elements combine in whole number proportions to make molecules, each of which has a characteristic combination of atoms of particular elements.

The development of chemistry seemed swift after the modern concepts of elements and atomic theory took hold. Avogadro contributed his hypothesis about the proportionality of gas volumes and number of particles, an idea that eventually turned into the mole concept. Wohler synthesized urea (an organic compound) from inorganic sources, throwing the theory of vitalism into crisis and ultimately founding modern organic synthesis. Chemical industries developed around the world, fueled by a better understanding of the structure and behavior of matter.

The final step, as far as we're concerned here, was the realization that atoms themselves were not indestructible. (You may recognize this as a modification of one of Dalton's original ideas about atoms.) By 1880, scientists like William Crookes reported on extensive investigations of Geissler tubes, which were high-quality (for the time) vacuum discharge tubes with small amounts of gaseous materials in them. Under certain circumstances, the discharges would emit radiation that would cause other materials like zinc sulfide to glow, or *fluoresce*. (See Figure 1.2.) Experiments suggested that this radiation, called *cathode rays*, had an electric charge. Conflicting reports and hypotheses led to detailed analyses of the phenomenon by J. J. Thomson. In 1897, Thomson presented evidence that cathode rays were

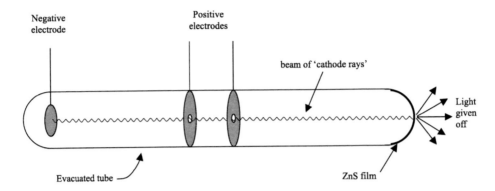

Figure 1.2 Electrodes inside a (mostly) evacuated tube form a discharge when a voltage is applied. Holes in the positive electrodes encourage the formation of a collimated beam of "cathode rays." Among other things, the cathode rays induce a film of zinc sulfide to fluoresce where the rays strike the film. Scientists were able to establish that cathode rays were actually charged particles by subjecting the beam to electrical and magnetic fields.

composed of tiny electrically charged particles that were smaller than an atom. The name *electron* was given to the individual particle.

While this announcement met with much skepticism, other experiments supported Thomson's ideas. These other investigations culminated in the famous Millikan oil-drop experiment, performed between 1908 and 1917 and illustrated in Figure 1.3. This work established the absolute charge in an individual electron, and when that was combined with the known charge-to-mass ratio (which was determined using magnetic fields), it verified that an electron was only 1/1837 the size of a hydrogen atom. Atoms, then, were not indivisible, but were instead composed of tinier parts.

The discovery of the proton, another subatomic particle that was positively charged, followed not long after. The existence of the neutral neutron was not verified until 1932. The arrangement of protons and electrons (and later, neutrons) in atoms was debated until 1911, when Rutherford postulated the nuclear atom. Based on experiments of sending α particles from radioactive materials toward a thin metal foil (Figure 1.4), Rutherford suggested that most of the mass of the atoms (protons and, eventually, neutrons) was concentrated in a central *nucleus* while the relatively light electrons occupied the space around the very spatially tiny nucleus.

The general view of matter as nuclear atoms has changed little since Rutherford's ideas. The *behavior* of such atoms has undergone some dramatic shifts in understanding, as our ability to measure such behavior has changed over time. Spectroscopy has always been at the center of our abil-

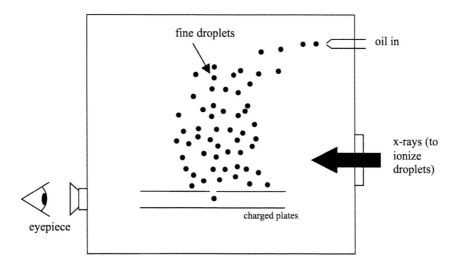

Figure 1.3 A diagram of Millikan's oil drop experiment. Oil droplets are generated by an atomizer and injected into a chamber. Here they are exposed to x rays, which ionize some droplets. Occasionally an ionized droplet falls between two charged plates, and the experimenter can vary the charge on the plate to see what charge is necessary to levitate the droplet. By making measurements on hundreds of droplets, Millikan determined that the magnitude on the charged droplets were all multiples of ~1.6×10^{-19} coulombs. This was how the fundamental charge on the electron was determined.

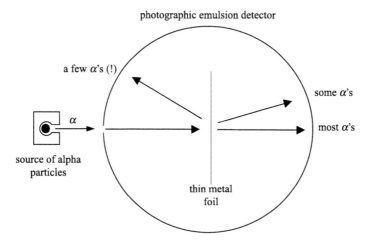

Figure 1.4 A diagram of Rutherford's experiment on the structure of the atom. Alpha particles from a radioactive source are directed toward a very thin metal foil. Most alpha particles passed right through the foil. Some are deflected a few degrees to one side. A very few were—surprisingly—deflected back toward the source! These results were interpreted in terms of a nuclear atom, with the protons (and later neutrons) in a tiny central nucleus and the electrons in orbit about the nucleus.

ity to measure behavior at the atomic and molecular level. But before we can discuss that topic, we turn now to the basic tool we use to study matter.

1.3 Light

What is light? Interestingly, throughout history this question did not seem to generate much speculation. Light was deemed to be either something that objects emitted so that we could see them, or something that was emitted by our eyes and bounced off objects. The basic behavior of light—it reflects, it refracts, it comes in colors, you can make various optical components like mirrors and lenses and prisms to manipulate it—became well understood, but that seemed to be the extent of the formal investigation of light. There were some attempts at increased understanding, notably by Claudius Ptolemy (first century A.D.), Abu Ali al-Hasan ibn al-Haytham (tenth century A.D.), and Robert Grosseteste and his student Roger Bacon (twelfth and thirteenth centuries A.D.), but they were apparently more phenomenological rather than theoretical, and little progress was made.

Until the seventeenth century, at least. In 1621, Snell discovered his law of refraction (which was not published until 1703), and Pierre de Fermat discovered the principle of least time and used it to explain Snell's law of refraction. But the real battle over the nature of light began in the 1660s with Robert Hooke.

Hooke was an outstanding scientist who had had the historical misfortune of being overshadowed by contemporaries who became more famous (like Boyle, Halley, and Newton). For example, Hooke studied harmonic motion of oscillators, published a widely read book *Micrographia* in which he presented drawings of microscopic organisms and structures that he viewed through a microscope, and was an excellent experimentalist. (He constructed the vacuum pumps that Boyle used to vary gas pressure in his studies of gases.)

Hooke's work on light is noteworthy because he was apparently the first credible scientist to propose, in *Micrographia*, that light is a very fast *wave*. He suggested that light, like sound, is a *longitudinal* wave; this contrasts with water waves, which are *transverse* waves. (See Figure 1.5.) In the late 1670s, Dutch physicist Christiaan Huygens provided additional arguments that light is a wave.

The competing hypothesis on the nature of light was represented by Isaac Newton. Newton was the first to demonstrate that white light is made by the combination of various colored light. (Newton was the one who proposed the name *spectrum* for the ghostly band of colors formed when a slit of white light is passed through a prism.) Newton proposed that light is composed of corpuscles, tiny particles that travel in a straight

Figure 1.5 Longitudinal versus transverse waves. Hooke proposed that light was a longitudinal wave. In this sort of wave, the medium is alternately compressed and rarefied in the direction of motion, as suggested by the top diagram. Dark areas represent compressed media, light areas are rarefied media. Sound waves are longitudinal waves. The other type of wave is a transverse wave, in which the medium moves perpendicular to the direction of motion, as suggested by the bottom diagram. Water waves are transverse waves.

line, which was why light makes sharp shadows and does not curve around corners like sound and water waves do. Newton's corpuscular theory of light gained adherents in part because of his fame—another example of influence winning converts.

The issue was apparently settled in 1801 when English scientist Thomas Young performed his double-slit experiment, illustrated in Figure 1.6. When light is passed through a thin slit in a mask and the image is projected onto a screen. The screen shows an expected intensity pattern: a bright vertical center directly opposite the slit, with the brightness decreasing as you move away from the position directly opposite the slit [as shown in Figure 1.6(a)]. On this basis, one would think that if we had two slits, we would get two images with bright centers and decreasing intensity as you move away from the points directly opposite the slits. Instead, what you actually see is depicted in Figure 1.6(b). A series of alternately bright and dark regions, with the brightest region *in between* the two slits, and the bright regions off to either side getting less and less intense. Young argued that this demonstrated the known *interference* phenomenon of waves, proving that light must, therefore, be a wave. Since Young's experiment, the wave nature of light has not been seriously questioned. Whether light is a transverse wave or a longitudinal wave was still questionable, but there was no denying that light had wave properties. (Light is actually treated as if it were a transverse wave.)

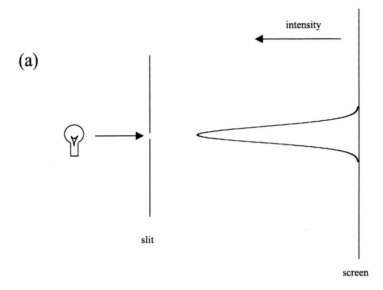

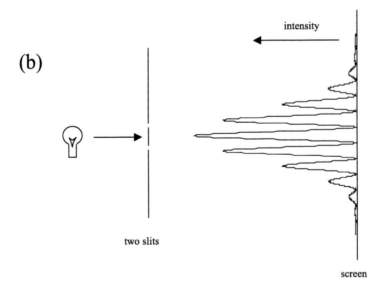

Figure 1.6 Young's double slit experiment. (a) When light passes through a single narrow slit, the intensity pattern of the image projected onto a screen shows a central bright region, with decreasing intensity seen on either side of the central bright region. (b) When light passes through two closely spaced slits, instead of a double image, there are interference fringes. Young used this as support of the idea that light is a wave.

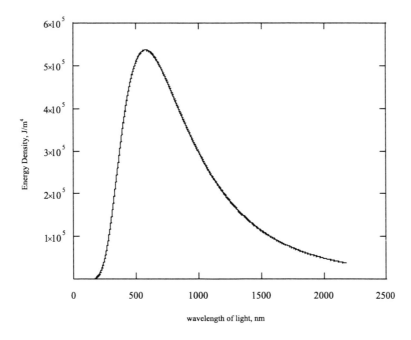

Figure 1.7 Intensity of light emitted from a blackbody versus wavelength. The temperature of the blackbody is 5000 K. Classical Science was not able to explain why blackbodies emitted light with this distribution.

However, this is not the end of the story. Further investigations into the behavior of light raised additional questions. In particular, the behavior of blackbodies was problematic. A blackbody is a perfect emitter or absorber of radiation. While nothing in the real world is perfect, very good approximations of blackbodies are easy to make (a small cavity with a tiny hole in it will suffice). You might think that a perfect emitter of light would emit the same amount of light at all wavelengths—but it does not. A blackbody emits light whose intensity depends on the temperature and wavelength in a complex way; a plot of the intensity of light emitted is shown in Figure 1.7. Scientists in the late nineteenth century were unable to explain this behavior. Perhaps the most successful attempt to explain the behavior of light in classical terms was the Rayleigh-Jeans law, which had the expression

$$\mathrm{d}\rho = \left(\frac{8\pi kT}{\lambda^4}\right)\mathrm{d}\lambda, \tag{1.1}$$

where $d\rho$ is the energy density of the emitted light (which is related to intensity), k is Boltzmann's constant, T is the absolute temperature, and $d\lambda$ is the wavelength interval. This expression matched experimental measurements in the long-wavelength region, but not in the short-wavelength region. In fact, the Rayleigh-Jeans law predicted an ever-increasing intensity as one goes to increasingly shorter wavelengths of light, approaching an infinite amount as the wavelength approaches the scale of x rays or gamma rays! This behavior was termed the ultraviolet catastrophe and clearly does not happen (else we would all be killed by the infinite amount of x rays being given off by matter). Approximations were also proposed (most successfully, by Wien) for the long-wavelength side of the maximum in Figure 1.7, but a single model eluded nineteenth-century scientists.

In December 1900, German physicist Max Planck proposed an expression that fit the entire plot, not just one side. Planck reasoned that since light was interacting with matter, matter itself must be behaving like little oscillators. Planck proposed that these oscillators couldn't have any arbitrary energy, but instead has a specific energy E that is related to the frequency v of the oscillation:

$$E = hv, \tag{1.2}$$

where h is a proportionality constant now known as *Planck's constant*. By making this assumption and using some thermodynamic arguments, Planck derived the following expression for the energy density:

$$d\rho = \frac{8\pi hc}{\lambda^5}\left(\frac{1}{e^{hc/\lambda kt} - 1}\right)d\lambda. \tag{1.3}$$

The variables in Eq. (1.3) have their normal meanings. A plot of this expression looks almost exactly like the experimental plots of blackbody radiation, suggesting that Planck's assumptions has some validity.

Some scientists, however, dismissed Planck's work as mere mathematical games with no value other than to predict a curve. There were questions about whether there was any real physical meaning to Planck's proposed relationship between energy and frequency. In 1905, however, Albert Einstein gave Planck's proposal more direct experimental support. Einstein applied Planck's equation $E = hv$ to light itself by suggesting that light of a particular frequency has a particular energy, in accordance with Planck's equation. Einstein then used this to explain the photoelectric effect, in which metals can emit electrons when certain wavelengths of light are shined on their surfaces. Thus, Einstein ultimately argued that light acts like a particle of energy, and the word "photon" was eventually coined by G. N. Lewis to describe a "particle" of light.

Additionally, in 1923, Arthur Compton showed that the scattering of monochromatic (i.e., "single color") x rays by graphite resulted in some of the x rays being shifted to a slightly longer wavelength. Compton used this evidence to argue that photons have *momentum* in addition to energy.

What type of material has specific energy *and* momentum? Why, particles, of course. Thus there is ample evidence to support the idea that light is acting like a particle (and thereby exonerating Newton).

Is light a particle, or light a wave? While some use the term "wavicle" or speak of "wave-particle duality," perhaps it is the question itself that is improper. In being described as having wavelength, frequency, interference behavior, and such, light is displaying wave properties. In having a certain specific (or *quantized*) energy and momentum, light is displaying particle properties. Light behaves as a wave or as a particle, depending on which property you are considering. Ultimately, it is limited thinking on our part to suggest that light *must* be either a particle or a wave, but not both.

1.4 Quantum Mechanics and Spectroscopy

The *quantum theory of light*, as proposed by Planck and interpreted by Einstein, completely changed how science deals with the molecular, atomic, and subatomic universe. This change in perspective is so profound that the year 1900, when Planck proposed his explanation of blackbody radiation, is typically considered the dividing line between Classical Science and Modern Science.

In the first 25 years of the twentieth century, there were several important advances. The nuclear structure of atoms was enunciated by Rutherford (see above), Bohr proposed a model of the hydrogen atom in which angular momentum was also quantized, and in 1923 Louis de Broglie proposed a relationship for the wavelength of a particle of matter (after all, if light could have particle properties, why can't particles have wave properties?):

$$\lambda = \frac{h}{p}, \tag{1.4}$$

where h is Planck's constant and p is the linear momentum of the particle. This set the stage for the development of quantum mechanics. After all, very small particles have a very small momentum, implying [because momentum is in the denominator of Eq. (1.4)] that they have a large

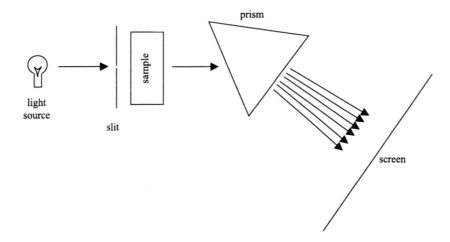

Figure 1.8 A simple schematic of Bunsen and Kirchhoff's spectroscope. Light from some source passes through a slit, a sample, and a prism before being projected onto a screen. Absorbed light is represented by dark lines superimposed on a rainbow-type spectrum. If the sample is heated and is emitting light, then the source is omitted and the spectrum consists of bright lines of light at wavelengths that are emitted by the sample. Bunsen and Kirchhoff showed that these wavelengths absorbed or emitted are characteristic of the elements in the sample.

"wavelength." If this is the case, then any understanding of the motion of very small particles must take wavelength behavior into account.

In 1925–26, Erwin Schrödinger and Werner Heisenberg developed independently a new model for electron behavior in atoms and molecules. Although they are mathematically equivalent models for all practical purposes, the majority of scientists use Schrödinger's formalisms. Schrödinger postulated that a system's behavior is contained in an expression called a *wavefunction*. Wavefunctions must have certain mathematical restrictions and must satisfy a partial differential equation, called the Schrödinger equation, that naturally yields the energy of the system. For most physically relevant systems, the energy of the system ends up being quantized, i.e., it has a specific value. In particular, the Schrödinger equation could be used to predict the spectrum of the hydrogen atom, and could also be applied to understand the spectra of other atoms and molecules—but we have gotten a bit ahead of ourselves.

Scientists have been studying the spectrum of light since Newton demonstrated its existence. In 1802, William Wollaston—followed in 1814 by Joseph Fraunhofer—noted some dark lines in the spectrum of the sun, thus unknowingly founding spectroscopic analysis. Circa 1859–60, German scientists Robert Bunsen and Gustav Kirchhoff invented the spectroscope, a device to systematically study a spectrum. A diagram of a simple spectro-scope is shown in Figure 1.8. Light from a source is passed through a sam-ple and then through a prism (or, alternately, through a rotatable prism and then through a sample), and then projected onto a screen. Alternatively, a sample could be heated to high temperature and the light emitted by the sample would be passed through a prism and then projected onto a screen. By observing a variety of samples with the spectroscope, Bunsen and Kirchhoff were able to show that each element contributed a *characteristic series* of absorbed wavelengths of light (for samples absorbing light) or, when light is given off by a heated sample, a characteristic set of wave-lengths of light are given off. Thus, Bunsen and Kirchhoff invented spec-troscopy as a method of determining what elements are present in a sample.

In short order, Bunsen and Kirchhoff identified two new elements, rubidium and cesium (both names deriving from the color of a very bright line in their respective emission spectrum, red for rubidium and blue for cesium; the element indium is also named after the bright indigo line in its emission spectrum). Thallium was discovered by its unique spectrum by Crookes in 1861, and its name derives from the Greek word *thallos*, mean-ing "green twig." Helium was detected spectroscopically on the sun in 1868 by Janssen, and finally discovered on earth by Ramsey in 1895. Samarium was also discovered spectroscopically, by Boisbaudran in 1879. Spectroscopy very quickly established its utility.

Spectroscopy was not confined to the visible region, however; it quickly spread to other regions of the electromagnetic spectrum. However, progress was delayed until photographic, instrumental, or—ultimately—electronic methods were developed to detect nonvisible photons. Modern spectros-copy spans virtually the entire electromagnetic spectrum.

In the development of ideas that ultimately led to the revolution of Modern Science, there was one issue that was intimately related to spec-troscopy: exactly *why* do atoms give off or absorb light that has only certain specific wavelengths? Or in terms of the quantum theory of light, exactly why do atoms absorb or emit light of only certain energies?

Particularly curious was the spectrum of hydrogen. Its spectrum in the visible region consists of four lines, as represented in Figure 1.9. In 1885, Swiss mathematician J. J. Balmer showed that these lines fit the following formula:

$$\frac{1}{\lambda} = R\left(\frac{1}{4} - \frac{1}{n^2}\right), \tag{1.5}$$

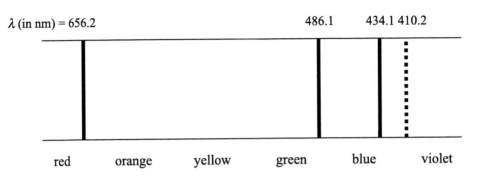

λ (in nm) = 656.2　　　　　　　　　486.1　　　434.1 410.2

red　　　　orange　　　yellow　　　green　　　blue　　　violet

Figure 1.9 A representation of the visible emission spectrum of H. The last line is dotted because it is sometimes difficult to see. Additional lines in this series are in the ultraviolet, and so not visible. There are other series of lines in the infrared and ultraviolet and other regions of the electromagnetic spectrum. Classical Science failed to explain why H had such a simple spectrum, although simple formulae were proposed for the wavelengths of the lines of light.

where λ was the wavelength of the light emitted, R was a constant, and n was either 3, 4, 5, or 6. (It was shown later that n could be larger than 6, but then the line of light is in the not-visible ultraviolet region of the spectrum.) After looking in other regions of the spectrum, other series of lines were detected, and in 1890 Johannes Rydberg generalized Balmer's formula as

$$\frac{1}{\lambda} = R_H\left(\frac{1}{n_1^2} - \frac{1}{n_2^2}\right), \tag{1.6}$$

where n_1 and n_2 are integers that are characteristic of the series of lines of light. (The one that Balmer discovered is known as the Balmer series.) Because the constant R now applies to every series of lines for the hydrogen atom's spectrum, it is relabeled R_H and is called the *Rydberg constant*.

Why is the hydrogen atom's spectrum so simple and easy to model? Classical Science could not explain it, nor could it explain the more complex series of lines emitted by other, more complex atoms. It was not until Bohr's theory of the hydrogen atom in 1913 that hydrogen's spectrum could be explained on theoretical grounds (by proposing that angular momentum, like energy, can also be quantized). However, Bohr's explanation was limited—and incorrect in some of its details. It was not until the development of quantum mechanics in the 1920s that a satisfactory theoretical framework for spectroscopy was developed, i.e., quantum mechanics. By then, spectroscopy—using different regions of the electromagnetic spectrum—was showing that energy levels of atoms and molecules were even more complex than was initially realized. But, quantum mechanics could still be used to explain the increasingly complex picture of atomic and molecular behavior.

Spectroscopy, then, was not only integral in the development of quantum mechanics—still our best theory of atomic and molecular behavior to date—but is inherently a quantum-mechanically based phenomenon. An adequate description of quantum mechanics is beyond the scope of this book. Indeed, some might argue that a "tutorial on quantum mechanics" is impossible! The important point is to realize is that spectroscopy is ultimately explained by quantum mechanics. In fact, whether you realize it or not, when you measure a spectrum, you are performing applied quantum mechanics!

References

1. I. Asimov, *Asimov's New Guide to Science*, Basic Books, Inc., New York (1984).
2. D. R. Lide, *CRC Handbook of Chemistry and Physics*, 76th ed., CRC Press, Boca Raton, FL (2000).
3. D. W. Ball, *Physical Chemistry*, Brooks-Cole Publishing Company, Forest Grove, CA (2003).

Chapter 2
LIGHT AND ITS INTERACTIONS

2.1 Properties of Light Waves

Under most conditions, light acts like a wave. According to Maxwell's equations, light is composed of oscillating electric and magnetic fields that pass through a vacuum at a certain constant velocity, c. The electric fields and magnetic fields are perpendicular to each other, and both are perpendicular to the direction of travel (see Figure 2.1). The electric and magnetic fields have specific directions as well as magnitudes, and so are properly thought of as vectors. It is interesting to note that two aspects of light have particle-like behavior: its energy (a fact deduced by Max Planck) and its momentum (first observed by Arthur Compton in 1923).

Like anything that acts as a wave, the behavior of light can be described by mathematical equations. Perhaps the simplest way to describe the magnitude of the electric (**E**) and magnetic (**B**) fields is by using a general equation in terms of the sine function:

$$E = A_e \sin\left(\frac{2\pi z}{\lambda} - 2\pi vt\right), \tag{2.1}$$

$$B = A_b \sin\left(\frac{2\pi z}{\lambda} - 2\pi vt\right), \tag{2.2}$$

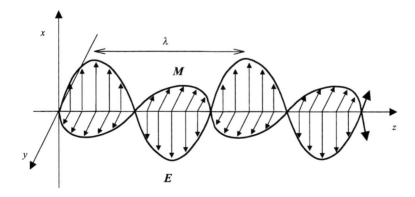

Figure 2.1 The wave representation of light. The axis of propagation is the positive z axis. The electric field vector **E** and the magnetic field vector **B** are shown (not necessarily to scale) along with the wavelength λ.

where A_e and A_b are the amplitudes of the electric and magnetic fields, respectively; λ is the wavelength of the light; and v is the frequency of the light. The variable z represents the position along the propagation axis (where we assume that the light is moving along the arbitrarily designated z axis), and t is time (that is, the field varies in time). If there is another (usually) constant term *inside* the sine function, this term implies that the magnitudes of the electric and magnetic vectors are not zero when z and t equal 0; we refer to this as a nonzero phase.

Theoretically, the ratio of the magnitudes A_e/A_b is equal to c, so the electric field has a much larger amplitude than the magnetic field. Equations (2.1) and (2.2) relate how the amplitude of the electric and magnetic fields vary with time t and distance z. Like any wave, the velocity of light, represented as c, is equal to its frequency times its wavelength:

$$c = \lambda \cdot v. \tag{2.3}$$

Typically, absorption and emission spectroscopy depend on an ability to measure the intensity of the light waves, because for light waves, intensity is related to amplitude. (For classical waves, amplitude is related to energy. Planck deduced that the energy of light was related to frequency. All forms of spectroscopy require an ability to differentiate between the differing energies of light.)

The above equations did not address the vector property of the electric and magnetic fields; in fact, the equations are more properly written as

$$\mathbf{E} = \mathbf{A_e}\sin\left(\frac{2\pi z}{\lambda} - 2\pi vt\right) \tag{2.4}$$

$$\mathbf{B} = \mathbf{A_b}\sin\left(\frac{2\pi z}{\lambda} - 2\pi vt\right), \tag{2.5}$$

where $\mathbf{A_e}$ and $\mathbf{A_b}$ are vector amplitudes and indicate the specific direction of the fields.

The vector amplitudes are very important because certain specialized forms of spectroscopy depend on the exact direction of the fields; that is, certain techniques are more dependent on the direction of the field vector than on its magnitude. This is called polarization spectroscopy. Because propagation is along the z axis and the fields are perpendicular to the x axis, the vector amplitudes can only exist in the x and y directions; hence, the vector amplitudes can be written generally as

$$\mathbf{A} = A\pi = A(\pi_x \mathbf{i} + \pi_y \mathbf{j}), \qquad (2.6)$$

where A represents either the electric or magnetic amplitude, A is the scalar value of the amplitude, and π is a unit polarization vector, possibly complex, that describes the polarization properties of the light. The terms $\mathbf{i}$ and $\mathbf{j}$ are unit vectors in the x and y directions, and π_x and π_y are their relative magnitudes; the only constraint is that

$$|\pi_x|^2 + |\pi_y|^2 = 1. \qquad (2.7)$$

The relative magnitudes of the $\mathbf{i}$ and $\mathbf{j}$ vectors in A determine the polarization of the light. Let us consider the electric vector of a set of light waves (which is most common, giving its larger magnitude). If each light wave has its own characteristic values of π_x and π_y, it has no preferential vector direction, and the light is considered *unpolarized*. If for all waves $\pi_x = 1$ and $\pi_y = 0$, the amplitude vector would exist only in the x dimension:

$$\mathbf{A} = A\mathbf{i}. \qquad (2.8)$$

We would speak of this light as being polarized in the x direction. Because all waves are lined up in the same direction, we can also refer to this as *linear polarization*. Light waves can also be polarized in the y direction, which would correspond to $\pi_x = 0$ and $\pi_y = 1$. Z polarized light is not defined because the z direction is usually considered the direction of propagation.

Finally, consider what happens if π_x equals $1/\sqrt{2}$ and π_y equals $i1/\sqrt{2}$, where i is the square root of -1. In this case, the polarization vector π becomes

$$\pi = \left(\frac{1}{\sqrt{2}}\mathbf{i} + i\frac{1}{\sqrt{2}}\mathbf{j} \right). \qquad (2.9)$$

(Do not confuse the two i's in Eq. (2.9)!) The net result of this is to make the propagation vector a corkscrew or helical shape, which in a right-handed coordinate system is considered a left-handed helix. Light having this polarization vector is called *left circular polarized light*. If the two summed terms in Eq. (2.9) are subtracted instead of added, the light becomes *right circular polarized light*. If π_1 and π_2 have different but constant values, the wave is called *elliptically polarized*. The various polarizations are illustrated

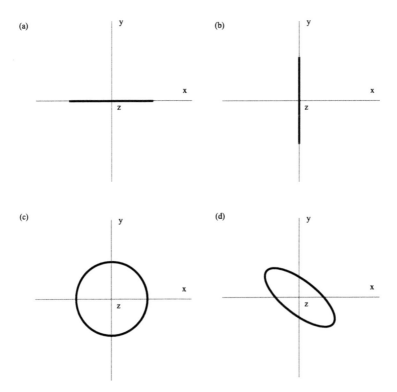

Figure 2.2 A representation of the various polarizations of light, seen by looking down the z axis of propagation. (a) x-polarized light. (b) y-polarized light. (c) circularly polarized light. (d) elliptically polarized light.

in Figure 2.2. For Figure 2.2(c) and (d), the specific polarization can be either clockwise (right circularly/elliptically polarized) or counterclockwise (left circularly front/elliptically polarized).

All of these properties of light waves—frequency, wavelength, phase, amplitude, intensity, energy, and polarization—are of interest to spectroscopists. When light interacts with matter, one or more of these properties of the light wave changes. If none of them changed, then spectroscopy would not be possible. As such, it is important for spectroscopists to realize which properties of light waves they are altering in order to have a better understanding of the spectroscopic technique.

2.2 Interactions of Light with Matter

Almost all of the knowledge we have about our universe is ultimately derived from the interaction of light with matter. Despite the complexity of the information sometimes generated by these interactions, the bottom

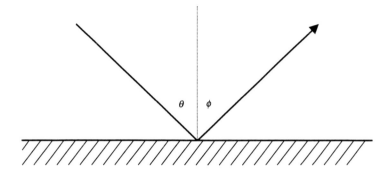

Figure 2.3 A simplified diagram of a light beam reflecting off a smooth surface. The dotted line perpendicular to the surface is called the normal; the incident angle with respect to the normal, θ, is equal in magnitude to the reflection angle with respect to the normal, ϕ.

line is that light interacts with matter in three ways: it can be reflected, it can be transmitted, or it can be absorbed. That is all. (We are ignoring scattering effects, which occurs for less than 1 in 10,000 photons. Scattering is actually an absorption-emission process anyway, so the previous statement is still accurate.) If one wanted to put this into an equation, then the original intensity of light, I_o, can be separated into three and only three parts:

$$I_o = I_r + I_t + I_a \, , \tag{2.10}$$

where the three parts are the intensities that are reflected, transmitted, and absorbed, respectively. The three processes usually occur simultaneously to some extent, so although we can discuss them independently, it should be kept in mind that all three processes are taking place at the same time.

2.2.1 Reflection

Reflection occurs when, to put it simply, a light wave bounces off the surface. Although the overall process can be interpreted as the complete absorption of a photon and then the re-emission of a photon of exactly the same wavelength and at a reflected angle equal to the incident angle, it is easiest to think of reflection simply as the photon bouncing off the surface. A diagram of a light ray reflecting off the surface is shown in Figure 2.3.

Reflection is not quite this simple, though. Although the angle of incidence equals the angle of reflection and the wavelength of the light is unaffected, the polarization properties of the reflected beam can be modified. Unpolarized light can be considered a random combination of light polarized perpendicular to the plane of the reflecting surface (s-polarized light) and light polarized parallel to the plane of the reflecting surface (p-polarized light). The polarization properties of reflected light will be discussed in Section 2.2.4 below. However, because a surface has different reflectivities for the two polarizations of light, the reflected beam may have different polarization properties from the incident beam. This property is especially important for surfaces that can also transmit at the particular wavelength of light.

2.2.2 Transmission

All materials transmit some portion of the electromagnetic spectrum. Many materials are completely transparent to x rays, for example, whereas other materials absorb infrared light. The light is not unaffected, however. Although light travels at characteristic speed through a vacuum (and that speed is considered a fundamental constant of the universe), light travels at different speeds in different media. For example, light travels at approximately 2.25×10^8 m/s in water, much slower than its normal 3.00×10^8 m/s in vacuum.

As a consequence, when light enters the new medium at an angle (instead of head-on), its path bends somewhat; this idea is called *refraction* and was known (and apparently even measured) by the ancient Greek philosopher Ptolemy. It was quantified in 1621 by Snell, and is thus called *Snell's law*:

$$n_i \sin \theta = n_r \sin \phi , \tag{2.11}$$

where the angles are defined as in Figure 2.4. The *index of refraction n* is defined as the quotient of the speed of light in that medium divided by the speed of light in a vacuum (this way all indices of refraction are > 1):

$$n = \frac{c}{v} , \tag{2.12}$$

where v is the velocity of the light in that medium. Each phase has its own characteristic index of refraction; thus, we have n_i for the index of refraction of the incident medium, and n_r for the index of refraction of the refraction medium in Eq. (2.11). The different polarizations of light have different

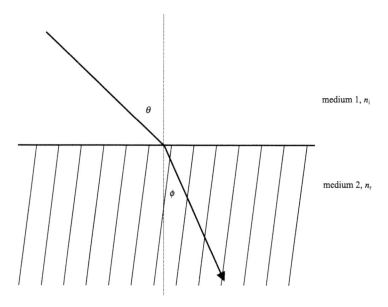

Figure 2.4 A diagram of the refraction of a light beam as it passes from one medium to another in obeyance of Snell's Law. In this example, $n_i < n_r$; if it were the other way around, ϕ would be greater than θ, not less than θ.

indices of refraction; the ns are also very dependent on the wavelength of light. (That is how a prism makes a spectrum or water vapor makes rainbows: the varying indices of refraction of the media cause a spatial separation in visible wavelengths of light.)

If one rearranges Snell's law, and assuming that the incident light is in the medium of higher index of refraction, for any two given media there is an angle of incidence in which the angle of reflection, ϕ, becomes 90 deg. At this point, the light is not transmitted into the medium but instead passes parallel to the surface, and larger angles of incidence are simply reflected off the surface of the "transmitting" medium. This angle, θ_c, is called the *critical angle,* and is given by the expression

$$\theta_c = \sin^{-1}\frac{n_r}{n_i}. \tag{2.13}$$

The critical angle is wavelength- and polarization-independent (although the indices of refraction may not be). At angles above the critical angle, light is not transmitted; it is reflected from the surface. It is this phenomenon that allows us to see into a lake very near the shore, but to see the sky reflected on the lake's surface farther from the shore. When the angle becomes greater than the critical angle, light is reflected from the lake's surface instead of being transmitted to illuminate what is under the water. (For water, the critical angle is approximately 48.8 deg.)

Though most people are unaware of its occurrence, light beams are always refracted whenever they pass into another medium, such as from air into glass. Although this sometimes introduces a distortion when making an observation—for example, looking out a window—usually the effect is so small that we can ignore it. The effect of refraction is more obvious when the interface between the media is curved, leading to an optical effect known as *lensing*.

2.2.3 Absorption

At certain wavelengths, a normally transmitting medium will absorb light. In such an instance, the photon is "destroyed" and its energy is converted into an atomic or molecular process—an electron changes orbit, a molecule vibrates or rotates differently, or some other effect. Usually, the wavelength(s) of the light absorbed is/are characteristic of the absorbing species, or *chromophore*. This is where the true power of spectroscopy lies, in that it imparts the ability to differentiate between different forms of matter because each has a unique spectrum of absorbed light. For most simple spectroscopic processes, two different energy levels are involved such that the difference between the energy levels, ΔE, is related to the frequency of the absorbed light by Bohr's frequency condition:

$$\Delta E = h\nu, \qquad (2.14)$$

where h is Planck's constant and ν is the frequency of the absorbed light. In the case of the absorbing medium, the index of refraction is complex and is rewritten as a wavelength-dependent complex expression

$$\hat{n} = n - i\kappa, \qquad (2.15)$$

where $\hat{n}$ is the complex index of refraction and κ the attenuation factor related to an *absorption coefficient*, α. The absorption coefficients are the core of spectroscopy: they relate how much light is absorbed each wavelength. The basic expression relating the intensity of light absorbed to the absorption coefficient is a very simple one, and is usually referred to as the *Beer-Lambert law*:

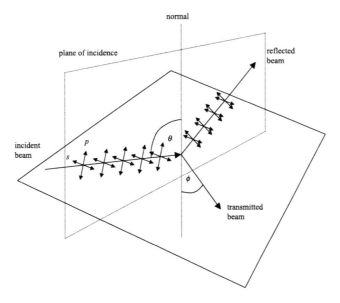

Figure 2.5 Definitions of s- and p-polarized light, which are defined in terms of the plane of incidence, not the plane of reflection. The transmitted beam (if present) also has s- and p-polarizations; they are omitted for clarity.

$$\ln\frac{I_0}{I} = \alpha c l,$$ (2.16)

where I_0 is the original light intensity, I is the intensity of the light after it passes through the absorbing medium, c the concentration of the absorbing species (not to be confused with the speed of light!), and l is the length of the absorbing medium. The Beer-Lambert law, or simply Beer's law, forms the basis of much of spectroscopy.

2.2.4 Polarization

As mentioned earlier, randomly polarized light can be described in terms of polarization perpendicular to a reflecting surface, and polarization parallel to a reflecting surface. Figure 2.5 shows how these two directions are defined. The plane of incidence is that plane marked out by the path of the incoming and reflected beams and that is perpendicular to the reflecting surface. It is with respect to this plane that s- and p-polarizations are defined. s-polarization is the component of the electric field that is perpendicular to this plane, and p-polarization is the component of the electric field that is parallel to this plane.

In cases wherein the reflecting surface can also transmit the light, reflection and transmission will both occur. If no absorption of light occurs, it is possible to quantify the amount of light reflected versus the amount transmitted. Let us assume for this discussion that light is traveling from a rare medium (for example, air) to a dense medium (for example, glass), or into a medium that has a higher index of refraction. This situation describes *external reflection*. It was found that the amount of light reflected depends on four things: the two indices of refraction, the angle of incidence (note that for a non-normal angle of incidence, the transmitted beam will be refracted and the angle of refraction can be determined from the indices of refraction and the angle of incidence using Snell's law), and polarization of the light. S-polarized light and p-polarized light reflect and transmit different proportions of their amplitudes. The reflected amplitudes can be expressed by the following equations:

$$I_{\parallel} = -\frac{\sin(\phi - \theta)}{\sin(\phi + \theta)} \qquad (2.17)$$

$$I_{\perp} = -\frac{\tan(\phi - \theta)}{\tan(\phi + \theta)}. \qquad (2.18)$$

Equations (2.17) and (2.18) are called *Fresnel's equations*. $I_{\parallel}$ and $I_{\perp}$ refer to s- and p-polarization, respectively. θ and ϕ are the angles of incidence and refraction, respectively. Because the angles are related by Snell's law the equations could have been written in terms of the indices of refraction and an inverse sine function, but that would have gotten messy.

For any two given indices of refraction, the fraction of the reflected power of s- and p-polarization (which is proportional to the square of the amplitudes) being reflected can be plotted versus the angle of incidence of θ. Such a plot is shown in Figure 2.6. The interesting point is that at a certain angle of incidence, the reflectivity of the p-polarized light is exactly 0; it is all transmitted through the denser medium. The angle at which this occurs, which is dependent on the indices of refraction of the two media, is called the *Brewster angle*. In this case where n_1 equals 1 and n_2 equals ~1.33, the Brewster angle is approximately 53.1 deg and can be shown to be equal to $\tan^{-1}(n_2/n_1)$. Windows tilted at the appropriate Brewster angle are used for gas lasers to induce a polarization on the laser beam. (Having a window at the Brewster angle also helps maximize laser throughput, but this is separate from the polarization issue.)

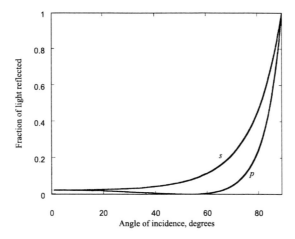

Figure 2.6 A plot of the power of s- and p-polarized light reflected vs. the angle of incidence. In this example, light is going from a rare medium (n = 1.00) into a denser medium (n = 1.33). At the Brewster angle of ~53.1 deg, the p-polarized light is completely transmitted, not reflected. These plots represent the squares of the Fresnel equations, Eqs. (2.17) and (2.18).

2.3 Transparent Media for Different Spectral Regions

One of the true revelations in spectroscopy is when you first learn that you should not use a UV-visible cuvette in an infrared spectrometer. Actually, you can—but it is not a good idea if you want a good spectrum across the mid-IR range (4000–200 cm^{-1}), the region of interest for much IR work. Do you know why? The material used to make sample holders for UV-visible spectrometers absorbs IR radiation between ~3000 and 150 cm^{-1}, making it virtually impossible to measure an IR spectrum of a sample. You can use these sample holders to measure spectra at higher or lower wave numbers, but you should not use them for the mid-IR region.

By the same token, you probably do not want to use KRS-5 for UV-visible spectroscopy. KRS-5 is a designation for a thallium bromide/thallium iodide composite that is used in IR spectroscopy (and used very carefully, because thallium compounds are poisonous). It is also a nice red-orange color, which is bound to disturb any attempts to measure a visible spectrum of a compound. The same is true for zinc selenide, ZnSe: it transmits IR light, but because it is a nice orange color, its applicability in visible spectroscopy is limited.

What is the point? Different materials are transparent to different wavelengths of light, and when performing the various types of spectroscopy, you must use the appropriate material if you need a sample holder, window, or lens. Furthermore, materials that are useful for one type of spectroscopy are not necessarily useful for other types of spectroscopy. This is a general statement, but not a global one. Windows of transparency exist for most materials in all spectral regions, but such windows do not generally make these materials useful for certain spectroscopic methods. For example, water absorbs IR radiation fairly well, but there are some windows where IR light is not absorbed. Thus, water has a limited use as a solvent when measuring IR spectra.

Sometimes materials are transparent to particular wavelengths of light for certain reasons, so it is useful to discuss these reasons. Perhaps the most obvious example is in visible spectroscopy. Sample holders, windows, lenses, and other optical components for visible spectroscopy are clear. That is, they do not absorb visible light, so the sample of interest should be the material absorbing visible light. Thus, most glass and plastic can be used in visible spectroscopy. If you want to extend the spectral range into the near-UV region, you should use a material that does not absorb UV light, either. Quartz (crystalline SiO_2, and sometimes called silica) is usually the material of choices; and good, high-quality silica can have a practical cutoff of 160 nm. (To give you an idea of how far into the UV region that is, the UV region starts at about 350 nm.) Sapphire, which is crystalline Al_2O_3, is also used in UV-visible spectroscopic applications. These notes about materials for UV-visible spectra are applicable to any spectroscopic technique that involves visible or ultraviolet photons, like fluorescence and phosphorescence spectra, laser spectroscopy, and so forth.

Molecular compounds have vibrational absorptions. Molecular compounds are not commonly used in IR spectroscopy as windows, lenses, etc., because those materials do not absorb IR light. Instead, simple ionic compounds are used to make components for IR spectroscopy. Sodium chloride (NaCl) is a very common IR material, as is potassium bromide (KBr). Both these materials are clear in their crystalline form. As mentioned above, KRS-5 and ZnSe are colored materials, so novice spectroscopists may blanch at using these materials to measure a spectrum. But even though they absorb visible light (which is why they are colored), they do not absorb IR radiation! They are transparent to IR radiation, so they can be used to measure the IR spectra of other materials. Cesium iodide (CsI) is transparent down to ~200 cm^{-1}, and is used for some far-IR work, although it is soft (and therefore easily scratched), somewhat hydroscopic (so it gets cloudy when exposed to humidity), and a little on the expensive side. Believe it or not, polyethylene can be used for far-IR spectra, with the exception of the region around 700 cm^{-1}, where polyethylene has a very strong absorption.

Curiously, diamond is a good IR transparent material. With the exception of an absorption at ~1280 cm^{-1}, diamond is transparent to IR light down to ~125 cm^{-1}. Because of diamond's other superior properties, some spectroscopists would love to use diamond optics more, but the expense can be prohibitive. Its use in spectroscopy is partially fueling research in diamond films by chemical vapor deposition (CVD) methods. (In 1975, the Soviet spacecraft *Venera 9* landed on Venus; some of its optical instrumentation had diamond optics.)

Beryllium windows are common on x-ray sources and detectors. As the lightest air- or water-insensitive metal, beryllium is transparent to a large range of x rays and so serves as a good window material.

For magnetic resonance spectroscopies, the idea is similar but the tactic is different—you want a sample holder that does not have a spectral signature in the region of interest, but in this case the spectrum is caused by magnetic resonance phenomena (either from unpaired electrons or certain nuclei). High-quality quartz tubes allow you to measure the spectrum of the sample itself, not the sample holder.

Many standard references on spectroscopy contain tables or diagrams listing regions of transparency for various materials. Interested readers are urged to consult such references.

References

1. R. S. Drago, *Physical Methods for Chemists*, 2nd ed., Saunders College Publishing, Philadelphia, 1992.
2. A. J. Gordon, R. A. Ford, *The Chemist's Companion*, John Wiley & Sons, New York, 1972.

Chapter 3
SPECTROMETERS

3.1 Introduction

In this chapter, we will consider the spectrometer itself. The problem is that no single description of a spectrometer exists. Spectrometers are designed to take advantage of the unique way that different types of light interact with matter. By the same token, most spectrometers have certain common elements. We will consider those elements.

First we will differentiate between the several general classes of spectrometers. Those that deal with absorption and emission phenomena will be considered first. Fourier transform (FT) spectrometers, including resonance spectrometers, will be considered next. Magnetic resonance spectrometers, which use magnetic fields simultaneously with electromagnetic light, will be discussed, as will the marriage of FT with magnetic resonance spectrometers.

3.2 Emission and Absorption Spectrometers

Energetically, emission and absorption are opposite processes: in absorption, a photon is taken in by an atom or molecule and causes a process; in emission, a process occurs and produces a photon. (In some forms of spectroscopy, such as Raman spectroscopy, the processes occur together.) In both cases, the important factors to consider are the energy of the photon *and* how many photons of each energy are involved. This second quantity is the *intensity*.

Because the same two things are important in both cases, spectrometers for emission and absorption spectroscopy are largely made up of similar components, but in a different order or orientation. For emission or absorption spectroscopy, the following components are necessary: a source of energy or photons, a method of energy differentiation (more about that soon), a sample that absorbs or emits photons, and a detector. Optics are also usually used to manipulate the photons. The source can be a light bulb, a laser, a magnetron, a synchrotron, electricity, a flame, or a hot ceramic rod. Detectors can be a simple heat absorber, photographic film, or light-sensitive electronics. Samples can be anything.

The "method of energy differentiation" is the key; normally it is performed by a *monochromator*. Devices such as colored filters, prisms (made of quartz, glass, or a salt), or gratings (etched in glass or metal or deposited

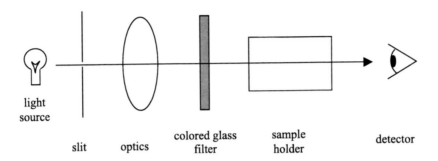

light
source

slit   optics   colored glass   sample   detector
                         filter   holder

Figure 3.1 A simple schematic of a colorimeter. Light passes through focusing optics and a colored filter before passing through a sample. A detector measures the intensity of colored light not absorbed by the sample.

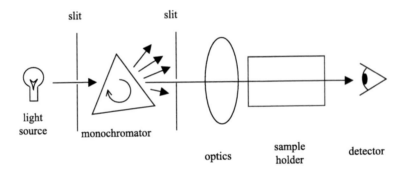

slit   slit

light
source   monochromator

                    optics   sample   detector
                               holder

Figure 3.2 A simple schematic of a dispersive absorption spectrometer. All dispersive spectrometers have a monochromator, either a prism or grating that disperses white light into its individual constituent wavelengths. Usually, the monochromator can rotate to allow light of different wavelengths to pass through the sample. Slits on either side of the monochromator allow only a narrow range of wavelengths of light to pass through a sample at any given time.

holographically) can separate electromagnetic light by frequency. This ability to separate light is *the* central part of spectroscopy. This is what allows us to determine what energies of light are absorbed or emitted by a particular sample.

The other components of absorption/emission spectrometers can take many forms and each is worthy of its own section. We will not deal with each component in detail here; rather, we will consider how these components make up a spectrometer.

A *colorimeter*, shown in Figure 3.1, has a source of photons, a monochromator with a filter that lets only certain wavelengths of light through, a

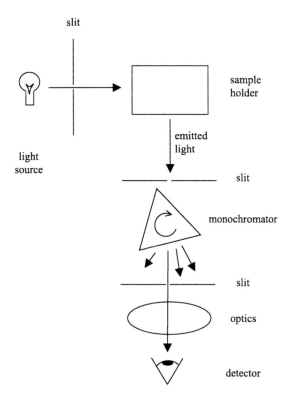

Figure 3.3 A simple schematic of an emission spectrometer. In this case, light given off by a sample is passed through a monochromator, then through a slit that allows only certain wavelengths of light to pass on to a detector. In this way, one can measure what wavelengths of light are emitted by a sample. Other means of sample excitation may be used—like electrical excitation or by heating the sample—instead of a light source.

sample, and a detector. Detectors simply measure how many photons reach them from the source. Measurements of light intensity with and without a sample indicate how many photons the sample absorbed.

A simple spectrophotometer is similar to a colorimeter, but has a variable wavelength monochromator and a prism or a grating instead of a filter. The instrument is illustrated in Figure 3.2. Although the monochromator allows only a small range of wavelengths (termed the *bandpass* of the spectrophotometer) to pass through a small slit and through the sample, mechanisms allow the monochromator to turn so that all wavelengths of light are swept or "scanned" across the slit and through the sample to the detector. One can manually select a wavelength of interest and measure the relative intensity of light reaching the detector through the sample; the classic "Spec 20" from the Milton Roy Company is a well-known example.

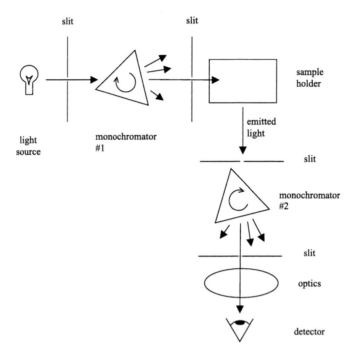

Figure 3.4 To select a particular wavelength of light to excite a sample, a first mono-chromator is placed before the sample. A second monochromator allows one to determine the emission spectrum.

The detector can be connected to the monochromator to electrically generate a plot showing intensity of light reaching the detector versus the position of a monochromator, or wavelength. This type of plot, intensity versus wavelength (or any other unit related to the energy of the photon), is called a *spectrum*. Machines that generate a spectrum are given the generic name *spectrometers*.

Emission spectrometers are essentially the same as shown in Figure 3.2. However, for emission spectrometers, the light of interest is not the light impinging on the sample, but light coming *from the sample*. In many instances, photons interact with a compound and then are re-emitted at a different wavelength (these interactions include fluorescence, phosphorescence, and Raman scattering). Because the wavelength of the emitted light must be determined, a monochromator is typically placed after the sample to differentiate the emitted light. This arrangement is shown in Figure 3.3. Note that the monochromator is usually placed out of line with the light from the sources. Emission usually occurs in all directions, but is best detected when light from the source is unlikely to interfere. This is especially true for laser-induced emission spectroscopies; care should be taken

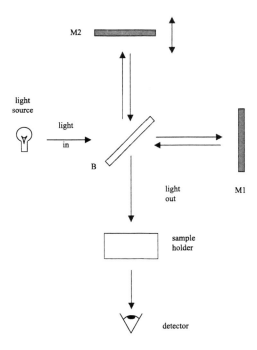

Figure 3.5 The basic components of an interferometric spectrometer. B = beamsplitter, M1 = fixed mirror, M2 = moving mirror. A laser beam (usually from a HeNe laser) travels parallel to the polychromatic light passing through the interferometer.

with residual light at the wavelength of the laser because it can be so intense that it burns out the detector.

Also, different wavelengths of incoming light promote different emissions from the sample. Thus, it is often useful to differentiate not only light emitted by the sample, but also the incident light from the source. This requires a dual monochromator system, as illustrated in Figure 3.4.

Most emission and absorption spectrometers are ultimately composed of the parts defined above, albeit with infinite variation. Such spectrometers are simple but very powerful tools to study matter.

3.3 Fourier Transform Spectrometers

The fundamental instrumental difference between a dispersive IR spectrometer and an FT-IR spectrometer is a group of optical components that compose an interferometer. The interferometer takes the place of a monochromator. A rudimentary sketch of an interferometer is shown in Figure 3.5. Simultaneously, light of all wavelengths from a source is split into two beams by a beamsplitter. Each beam is directed toward a mirror, one sta-

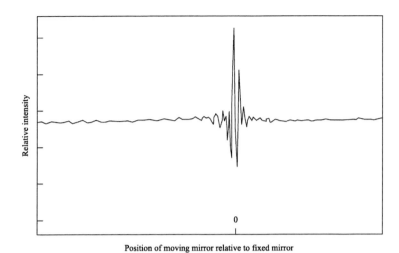

Position of moving mirror relative to fixed mirror

Figure 3.6 A typical interferogram. The point marked '0' is where both mirrors of the interferometer are the same distance from the beamsplitter, resulting in all constructive interference. Destructive interference at most other positions brings the intensity of the combined beam to some constant value. The scale on the x axis is usually cm (i.e. centimeters away from equidistance) or points (i.e. number of data points on either side of the centerburst).

tionary (M1) and one moving back and forth continuously (M2). After hitting the mirrors, the beams are reflected back, recombined at the beamsplitter, then travel through the sample to a detector. The interferometer has a unique (if not apocryphal) place in science history. It was invented by A. A. Michelson and used by Michelson and E. W. Morley in their seminal "ether drift" experiments.

Because an FT-IR spectrometer contains no monochromator, all wavelengths of light pass through the interferometer at the same time. But because light is a wave, constructive and destructive interference occurs when the two half-beams are recombined. In fact, because all wavelengths of light are traveling together, destructive interference dominates unless the mirrors are equidistant from the beamsplitter. The precise intensity of the recombined light depends on the relative position of the moving mirror. A plot of this intensity, called an *interferogram*, is illustrated in Figure 3.6. The tall part of the interferogram, called the *centerburst*, occurs when the distances of the two mirrors from the beamsplitter are equal. The overall light intensity drops off quickly due to destructive interference.

If the detector were set to measure the intensity of the recombined light beam versus the moving mirror position, the resulting signal would be an interferogram. But what is the relationship between an interferogram and a spectrum?

Lord Rayleigh was the first (in 1892) to show mathematically that the interferogram (a plot of intensity vs. distance or time) was related to a spectrum (a plot of intensity vs. inverse distance or inverse time) by a mathematical function called the Fourier transform. Generally speaking, if the intensity plot is described by some function of time, $h(t)$, then the Fourier transform of h is denoted $H(f)$ and is mathematically defined as

$$H(f) = \int_{-\infty}^{+\infty} h(t)e^{-i2\pi ft}\mathrm{d}t. \tag{3.1}$$

The new function $H(f)$ has units of frequency (inverse of time). The Fourier transform can be similarly defined for a function that depends on distance; upon performing the Fourier transform, the new function H has units of inverse distance. Given the mathematical relationship between imaginary exponential functions and sine and cosine functions, the Fourier transform can also be defined in terms of these functions. Sine and cosine Fourier transforms are equivalent to Eq. (3.1) above.

The Fourier transform spectrometer measures an intensity plot of light coming from a spectrometer in terms of the travel of the moving interferometer mirror M2. The intensity plot can be based on the distance the moving mirror travels or the time of each back-and-forth cycle of the mirror. Then, generating the Fourier transform of that plot calculates another plot of intensity, but in terms of inverse time (which spectroscopists call *frequency*) or inverse distance (which spectroscopists term *wavenumber*).Of course, an interferogram (see Figure 3.6) is not a function whose Fourier transform can be calculated easily. An interferogram is in fact a very complicated function, so although the ideas behind Fourier transform spectroscopy had been known since the late 1800s, it was only with the development of computers that FT spectroscopy became practical. A computer can digitize the interferogram and calculate what we would consider a spectrum. Only since the 1950s have FT instruments become laboratory fixtures.

Interferometric/FT techniques have many theoretical and practical advantages over regular dispersive ones. Books have been written on the subject, and interested readers are encouraged to see details elsewhere. One advantage is the use of a computer to store an interferogram, measuring several more interferograms and averaging them numerically to decrease the noise level of the computed spectrum. To do this, the computer must average the centerburst at exactly the right position along the mirror's path every time, otherwise this advantage is quickly lost and other artifacts appear in the spectrum. To achieve this regular positioning, the spectrometer needs a very good measuring device, and so all interferometers have a small laser associated with them (usually a helium-neon, HeNe, laser). The monochromatic beam from the laser passes through the interferometer just like the light from the source, and upon recombining produces interference fringes separated by exactly the same amount (632.8 nm for the red HeNe light). There is no centerburst on this recombined, monochromatic beam! These fringes act as a yardstick, allowing the moving mirror and a computer to stay synchronized.

There is one other complication. A Fourier transform spectrometer is inherently a *single-beam* spectrometer. All of the spectrometers in Figs. 3.1–3.4 can split the light beam at any point and send the beams down two identical optical paths, differing only in that one path contains a sample and the other does not. These would be *double-beam* spectrometers. A spectrum is generated by comparing the signal intensity of two detectors at identical wavelengths. Because of its unique optical path, it is difficult to split an interferogram signal and obtain a double-beam instrument. (It is not impossible; in fact, a commercial double-beam FTIR has been advertised recently.) Most FT spectrometers are operated in single-beam mode. First, a so-called *background spectrum* is measured and stored digitally. The background spectrum has everything in the optical path (sample holder, solvent, etc.) except the sample to be measured. Then, the sample is introduced and a second single-beam spectrum, the *sample spectrum*, is measured and stored digitally. The two single-beam spectra are then compared numerically to generate an absorption or transmission spectrum that we are typically familiar with.

Ultimately, the result is a computer-generated spectrum, exactly like one from a dispersive instrument. Other differences between FT and dispersive spectrometers are beyond the scope of this book. Common FT spectrometers are found for infrared, Raman, and magnetic resonance techniques, although the interferometer can be used in other spectral ranges (again the reasoning is beyond the scope of this book). However, when all is said and done, the Fourier transform spectrometer is a powerful tool for studying the interactions of light with matter.

3.4 Magnetic Resonance Spectrometers

Although magnetic resonance spectrometers are not technically a different type of spectrometer, there are some obvious differences in how they operate. Therefore, a separate section will be devoted to them. Magnetic resonance spectrometers take advantage of the effect of a magnetic field on certain samples.

All subatomic particles have a property known as *spin angular momentum*, or simply *spin*. The name is something of a misnomer because the particles are not really spinning on an axis. Quantum mechanics allows as quantized observables a *total* spin angular momentum (represented by the quantum number S for electrons and the quantum number I for nuclei) and a z component of the total spin angular momentum (represented by the quantum number m_S for electrons and the quantum number m_I for nuclei). For nuclei, values of m_I range from $-I$ to I in integer steps, allowing for $2I + 1$ possible values. Values of the angular momentum observables are related to the values of the quantum number, so it is convenient to simply refer to quantum numbers.

All electrons have a spin S of ½, and have two possible values of m_S, +½ and –½. The Pauli principle requires that two electrons in an orbital have opposite z-component spin (for example, m_S for one electron equals +½ and m_S for the other electron equals –½). Therefore, paired electrons have an overall zero net z-component spin. However, unpaired electrons—either as free radicals or as unpaired electrons in molecular orbitals (for example, in oxygen, O_2) or as unpaired electrons in degenerate atomic orbitals (for example, in transition metal compounds)—result in an overall nonzero net z-component spin. The same is true with nuclear particles: protons and neutrons individually have spins of ½. Together in a nucleus, the net z-component spin of the nucleus can be zero if all spins are properly paired, but *may* be nonzero (and different isotopes of the same elements will have different net nuclear z-component spins).

A species with a net z-component spin has a certain energy (E). This same species in the presence of a magnetic field (H) will have a different energy $E + E_H$, where E_H is the energy resulting from the interaction of the spin angular momentum and the magnetic field vectors. For nuclei, E_H can be calculated as

$$E_H = -\frac{m_I \mu \beta H}{I}, \tag{3.2}$$

where m_I and I are the quantum numbers from above, β is the nuclear magneton, and μ is the magnetic moment of the nucleus. H is the magnetic field strength, usually in units of gauss (G) or tesla (T). An analogous expression can be written for electrons in terms of S, m_S, the Bohr magneton, and the magnetic moment of the electron. For a given nucleus (or for electrons), I, m, and β are characteristic, and the possible values for m_I are dictated by I. Because m_I can have $2I + 1$ different values, the different states of the species having net z-component spin will have different values of E_H depend-

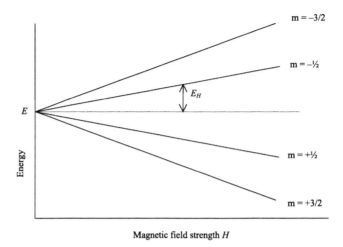

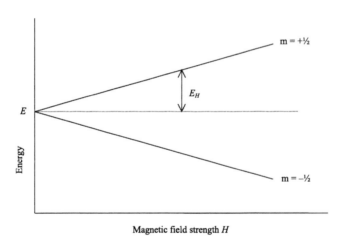

Figure 3.7 Energy level diagrams showing the change in energy of levels having different values of m for (a) a nucleus having $I = 3/2$, and (b) an electron. At $H = 0$, the energy of the particle is E. E_H is the change in energy due to interaction with the magnetic field.

ing on the m_I quantum number. Some states will go up in energy and some will go down, depending on the sign on m_I and charge on the species (negative for electrons, positive for nuclei). This behavior is illustrated in Figure 3.7.

So in a magnetic field, the $2I + 1$ different spin states of a nucleus having nonzero spin have different energies (and similarly for electrons). The difference in the energies of the states, ΔE, can be bridged with a photon if

$$\Delta E = h\nu , \qquad (3.3)$$

where ν is the frequency of the photon. Thus, by absorbing a photon, a species can go from one spin state to another spin state. Because ΔE (and therefore the frequency/wavelength/energy of the photon) depends on the strength of the magnetic field, there is a direct relationship between ν and H. This defines *magnetic resonance spectroscopy*. The two common magnetic resonance techniques are nuclear magnetic resonance (NMR) and electron spin (or paramagnetic) resonance (ESR or EPR) spectroscopy.

The idea behind the two spectroscopies is exactly the same. However, the practical manifestations of the techniques are not, due to the differences in β and μ in Eq. (3.3). (Table 3.1 lists values of m for various particles. The Bohr magneton and nuclear magneton have values of 9.274×10^{-16} erg/tesla and 5.051×10^{-18} erg/tesla, respectively.) In both techniques, samples are exposed to a magnetic field, typically from an electromagnet. The magnetic field strengths range from 1.4 to 14.1 T (or more) for NMR, depending on the nuclei of interest, and approximately 3400 G (0.34 T) for ESR. The electromagnetic radiation used is also different. In NMR, low-energy radio waves are used; whereas in ESR, higher-energy microwave radiation is appropriate.

Instrumentation for NMR and ESR is in theory the same as any other spectrometer, except for the addition of a magnet (usually an electromagnet). In practice, it is quite different. Because of the lack of useful monochromators in the microwave and radio regions of the spectrum, typically

Table 3.1 Spins and magnetic moments for various particles.

Particle*	I	Magnetic moment μ
e^-	1/2	~1.00115
^{1}H	1/2	2.79268
^{3}He	1/2	−2.1274
^{6}Li	1	0.82192
^{11}B	3/2	2.6880
^{13}C	1/2	0.702199
^{19}F	1/2	2.62727
^{31}P	1/2	1.1305
^{209}Bi	9/2	4.03896

*With the exception of the electron, all entries refer to the nucleus of the atom listed. (Source: Weast, R.C. *CRC Handbook of Chemistry and Physics*, 60th ed., Boca Raton, FL, 1979.)

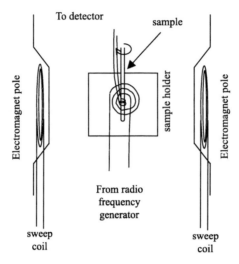

Figure 3.8 A diagram of an (absorbance-type) NMR spectrometer, showing a sample positioned between the poles of a magnet. The sample is spinning in the sample holder. The detector coils are supposed to be perpendicular to the rf field coils.

the frequency of radiation is kept as constant as possible, and the magnetic field strength is slowly varied. A detector (usually a balanced bridged electronic device) measuring the radiation output detects any absorption by a sample when the resonance condition is met:

$$hv = \frac{(\Delta m_I)\mu\beta H}{I}.\tag{3.4}$$

(Again, a similar expression can be written for electrons.) H and v are the only experimentally determined parameters. The possible values of Δm_I depend on the particles (electron or nucleus) under study.

Figure 3.8 shows a diagram of an (absorbance-type) NMR spectrometer. (A different type of NMR spectrometer will be discussed in the next section). The electromagnet has a set of "sweep" coils that alter the magnetic field strength by a small amount, typically a few hundred milligauss. The radio frequency (rf) source is a generator connected to coils set at right angles to the direction of the magnetic field. Linearly polarized rf waves are produced. The sample is usually placed in a slim tube and is spun to minimize the effects of inhomogeneities in the field. The detector—another coil wrapped around the sample—attaches to a bridge circuit. Using the vector properties of the polarized rf radiation, the detector notes the

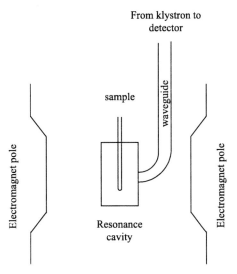

Figure 3.9 A diagram of an ESR spectrometer. The waveguide is typically behind the resonance cavity and not between the magnet poles (shown this way for clarity). The detector, not shown, is part of the klystron/waveguide system.

absorption of rf energy. The signal is then amplified and sent to a plotter or computer.

Figure 3.9 shows a diagram of the ESR spectrometer. Instrumentation is similar to NMR, except that the microwave radiation is generated by a *klystron* and the microwaves are delivered to the sample through a hollow rectangular tube called a *waveguide*. No sweep coils on the magnet are necessary because the magnet itself can perform the sweep. The sample holder is a resonance cavity in which the wavelength of the microwaves is such that a standing wave is established. The decrease in energy in the cavity resulting from absorption by the sample is detected by the imbalance of a bridge circuit. Although NMR spectra are plotted as absorptions, ESR spectra are plotted as derivatives to better see individual transitions.

Needless to say, actually operating an NMR or ESR spectrometer requires more background than is given here. Interpreting an NMR or ESR spectrum also requires more background! References at the end of this chapter give more information on these points.

3.5 Fourier Transform NMR

In Section 3.3, we introduced the idea of the Fourier transform and briefly described how it is applied to optical spectroscopy. In Section 3.4, we introduced magnetic resonance spectrometers. In that discussion, we described resonance spectrometers as if their magnetic fields scanned a range of values while detectors looked for absorption in microwave or radio-wave radiation as the energy difference between states varied with magnetic field. This is called the *slow-passage* or *sweep mode* for the spectrometer. However, magnetic resonance spectrometers can also operate using Fourier transform techniques.

Recall that nuclei having a nonzero net spin I (that is, $I \neq 0$) have a nuclear magnetic moment μ that is characteristic of the particular nucleus. (For hydrogen nuclei, $\mu = 2.7927$ nuclear magnetons, for example.) Classically, as opposed to quantum mechanically, the nucleus has a magnetic moment $\vec{\mu}$, which is a vector and is related to the total angular momentum $\vec{I}$ of the nucleus by the following equation:

$$\vec{\mu} = \gamma \cdot \vec{I}, \tag{3.5}$$

where γ is the *magnetogyric ratio*, which is also characteristic of the particular nucleus. When a nucleus is exposed to a magnetic field, the magnetic field vector $\vec{H}$ interacts with the magnetic moment vector $\vec{\mu}$ in such a way as to cause the magnetic moment vector to rotate, or precess, about the magnetic field vector. This precession is a change in direction of the magnetic moment vector, labeled $\dot{\vec{\mu}}$, and is given by the expression

$$\dot{\vec{\mu}} = \lambda \vec{H} \times \vec{\mu}, \tag{3.6}$$

where the $\times$ represents a vector cross product. The frequency at which the magnetic moment vector precesses is called the *Larmor frequency* ω and is given by γH, where H the magnitude of the applied magnetic field. These relationships are illustrated in Figure 3.10.[1]

In a bulk sample, all the nuclei's magnetic moment vectors contribute to an overall *net magnetization* of the sample, $\vec{M}$:

$$\vec{M} = \sum \vec{\mu}. \tag{3.7}$$

---

[1] In fact, the word "resonance" refers to the matching of the Larmor frequency and the applied radio frequency radiation.

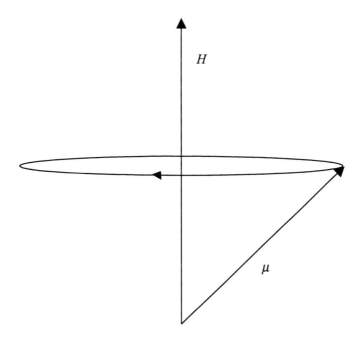

Figure 3.10 When a magnetic field *H* is applied to a magnetic moment m, the magnetic moment vector precesses about the magnetic field vector as shown.

In the presence of the magnetic field, the net magnetization precesses just like the individual magnetic moments. The expression for this is very similar to Eq. (3.6):

$$\dot{\vec{M}} = -\gamma \vec{H} \times \vec{M}. \tag{3.8}$$

In the presence of the magnetic field, some states originally having the same energy—called *degenerate* states—interact to different extents with the magnetic field and adopt slightly different energies, and electromagnetic radiation of the proper wavelength will cause a given nucleus to change states. This is an absorption and can be detected. How does this work? Consider a two-dimensional example of what happens to the net magnetization $\vec{M}$ when a nucleus absorbs energy and changes its state with respect to the magnetic field. Figure 3.11(a) shows four nuclei in their lowest energy state aligned with the magnetic field, and the net magnetization $\vec{M}$. Figure 3.11(b) shows the net magnetization when one of the nuclei changes state. Note that although the magnitude of the net magnetization

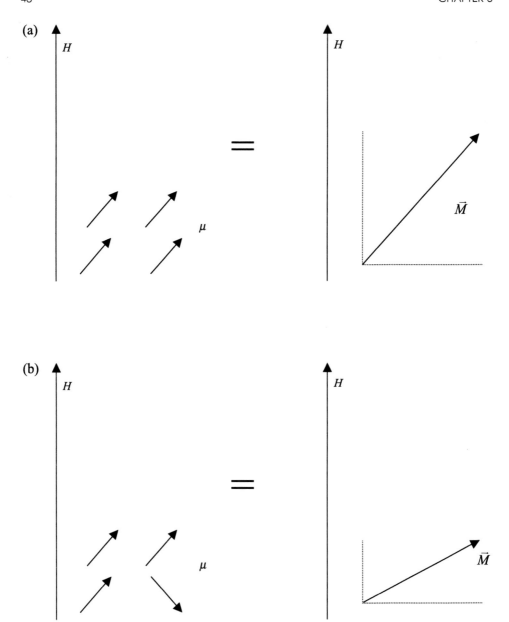

Figure 3.11 (a) A two-dimensional representation of four nuclei in a magnetic field. The dotted lines show the projection of the net magnetization vector along the imposed magnetic field, and perpendicular to it. (b) After one nuclei absorbs energy and now points in the opposing direction, the net magnetization vector has different projections (as shown by the changes in the lengths of the dotted lines). In real processes, the direction of the component of the magnetization vector is perpendicular to H (making it perpendicular to the plane of the page), and not parallel as shown.

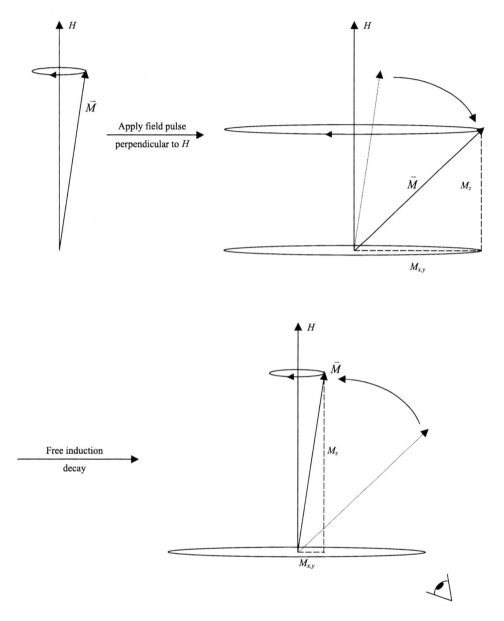

Figure 3.12 The changes in the net magnetization vector during the NMR transition process.

vector in one direction remains the same, the magnitude in the other direction, which is perpendicular to the field, has changed. (In the figure, the projection along the applied magnetic field is shown as in the plane of the

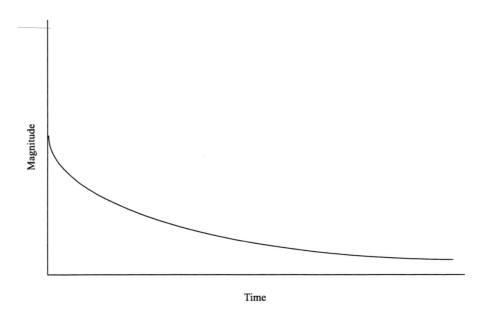

Figure 3.13 A simple free induction decay (FID) curve after a pulsed NMR signal. Most FIDs are more complex than this—see Figure 3.14.

paper; in fact, because of the right-hand rule, this particular projection would actually be out of the plane of the page and will cause the magnetization vector to rotate, as shown in Figure 3.10, but farther into the plane perpendicular to H.) This difference is detectable and provides the signal for NMR spectroscopy.

Most modern NMR spectrometers do not use a sweep method. Instead, they also take advantage of the Fourier transform. The process in three dimensions is illustrated in Figure 3.12. If a very short pulse ($<1/\omega$, where ω is the Larmor frequency) of a relatively strong magnetic field (approximately 100 G) were applied to the sample *perpendicular* to the first applied field, some of the nuclei will absorb energy and jump to a higher energy spin state, changing the magnitude of the projection of the net magnetization $\vec{M}$ perpendicular to the applied magnetic field. After the pulse, nuclei will slowly revert back to the lowest energy state over a period of time, realigning themselves with the applied magnetic field. A detector (represented by the eye in Figure 3.12) notes the decrease in the magnitude of the magnetization vector in the plane perpendicular to the applied (and static) magnetic field H. If the magnitude of the perpendicular projection is plotted over time starting immediately after the pulse, one would see a slow decrease in that magnitude. An example of what the magnitude might look

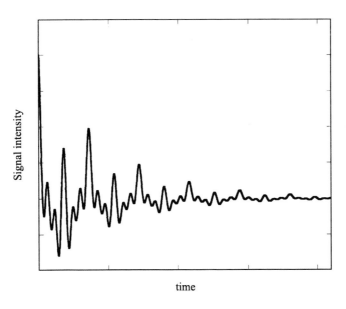

Figure 3.14 A more representative FID curve observed after an NMR pulse. The NMR spectrum is the Fourier transform of this curve.

like is shown in Figure 3.13 and is called a *free induction decay* (FID) curve. This figure shows a simple FID curve, where the nuclei are precessing about H at the Larmor frequency. Nuclei in different chemical environments will experience slightly different fields H_N, and so will precess at slightly different frequencies. Therefore, the FID will be a more complicated waveform, perhaps like that shown in Figure 3.14.

Figure 3.14 shows an FID curve plotted as intensity vs. time. The Fourier transform, applied to this waveform, converts the FID curve into an intensity vs. frequency plot; that is, a spectrum.

FT-NMR has the same FT-associated advantages as FTIRs: sampling all energies at once, measuring multiple spectra, and averaging them to reduce noise. FT-NMRs are also computer-driven, because the heavy calculation requirements make computers necessary. Once again, however, the Fourier transform shows itself to be important in another area of spectroscopy.

References

1. Ingle, J. D., Jr.; Crouch, S. R. *Spectrochemical Analysis*, Prentice-Hall, Englewood Cliffs, NJ, 1988.

2. Griffiths, P. R.; de Haseth, J. A. *Fourier Transform Infrared Spectroscopy,* Wiley-Interscience, New York, 1986.

Chapter 4
THE SPECTRUM

4.1 Introduction

A spectrum is a representation of what electromagnetic radiation is absorbed or emitted by a sample. The representation could be a plot, a diagram on a computer screen, even a list of wavelengths and intensities.

The word "spectrum" was coined by Isaac Newton, from a Latin word meaning "appearance." (The word "spectre," meaning ghost, shares the same root.) Newton projected his spectrum (plural *spectra*) on a wall or screen, and for a few hundred years the projection of a spectrum onto a surface was the best that could be done. (And even under such conditions, progress was made!) With the development of photography and, ultimately, electrical and electronic detection devices, a spectrum could be recorded permanently. Ultimately, the word "spectrum" came to represent the permanent record, rather than the dispersed white light (or other region of the electromagnetic spectrum).

In this chapter, we will treat a spectrum as the physical record of which wavelengths/frequencies/energies of light are absorbed or emitted by a sample. With that in mind, we will find that there are several popular ways of displaying a spectrum, although different types of spectroscopy typically display spectra in one or two common ways.

4.2 Types of Spectroscopy

Virtually all spectra show a plot of some signal versus a photon characteristic (energy, wavelength, wavenumber, frequency, etc.). The signal comes from the absorption or emission of a photon from an atom or molecule, and is accompanied by a concurrent change in state of the atom or molecule. Atomic or molecular states are dictated by wavefunctions (as discussed very briefly in the quantum mechanics section of Chapter 1). Technically, a single wavefunction defines the entire state of a molecule. However, it is a good approximation that the behavior of the electrons in a molecule can be considered independently from the behavior of the nuclei in a molecule, and they can be described by their own separate wavefunctions:

$$\Psi \approx \Psi_{el} \cdot \Psi_{nuc}, \qquad (4.1)$$

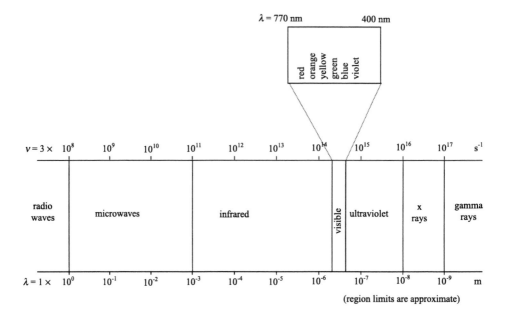

Figure 4.1 The electromagnetic spectrum. Wavelength and frequency boundaries are approximate.

where Ψ represents the wavefunction for the entire molecule, Ψ_{el} is the wavefunction of the electrons for a given set of nuclear coordinates, and Ψ_{nuc} is the wavefunction for the nuclei. This separation of electronic and nuclear wavefunctions is called the *Born-Oppenheimer approximation.*

What this approximation allows us usually to do is conveniently separate the various processes of a molecule, and spectroscopy takes advantage of that. Different types of spectroscopy typically involve different, *separate* molecular processes. (There are forms of spectroscopy that involve two (or more) processes simultaneously; except in some cases, we will not consider these in this book.) Further, different types of spectroscopy typically are limited to a specific range in the electromagnetic spectrum. This leads us to the very useful conclusion that the description of a certain type of spectroscopy can be (largely) indicated either by the molecular process involved, or the region of the electromagnetic spectrum used.

Figure 4.1 shows a stylized electromagnetic spectrum, while Table 4.1 lists the regions of the electromagnetic spectrum, the types of spectroscopies that use that region, and the process(es) involved. There are many other forms or types of spectroscopy than those listed in the Table, but it would be impossible to be all-inclusive. Table 4.1 includes the common absorption- or emission-based spectroscopic technique for that spec-

Table 4.1 Types of spectroscopy across the electromagnetic spectrum.

Region	Spectroscopy	Process Involved
Radio waves	Nuclear magnetic Resonance*	Changing nuclear spin orientation
Microwave	Electron spin Resonance	Changing electron Spin orientation
	(pure) Rotational	Changing molecular vibrational states
Infrared	Vibrational**	Changing molecular vibrational states
Ultraviolet	Electronic	Changing atomic or molecular electronic states
X ray	Inner electronic	Changing electronic states or ejecting electrons
Gamma ray	Mössbauer	Changing nuclear energy levels

*These types of spectroscopy require a magnetic field to differentiate between different spin states.
**Rotational transitions commonly superimpose themselves onto vibrational spectra.

tral region. Probably the biggest omission from this list is Raman spectroscopy, which is a differential type of spectroscopy (i.e., the spectrum is measured by differences in photon energy, rather than the energy of the photons themselves). Table 4.1 also omits various hybrid forms of spectroscopy, like photoacoustic spectroscopy, and does not include mass spectrometry because it can be argued that a mass spectrum is not a spectrum in the sense considered here.

Despite the different spectral regions and atomic or molecular processes probed, all (absorption-style) spectra have some common elements. In all cases, transitions between states are probed with dispersed light[1] of approximately the correct energy; and when the light has the same energy as the difference between two states, light is absorbed and the atom or molecule goes from one state to the next. Recall that this is written mathematically as

$$\Delta E = h\nu, \qquad\qquad (4.2)$$

---

[1] When we refer to "dispersed light," we mean that the light has been separated into a continuous band of constantly varying wavelength or frequency by some sort of monochromator, like a grating or a prism. Chapter 3 discusses these components of spectrometers.

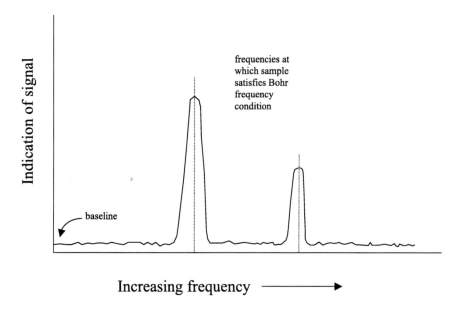

Figure 4.2 A model spectrum showing two spectroscopic signals moving out of the baseline.

where ΔE is the energy difference between the two states involved, h is Planck's constant ($= 6.626 \times 10^{-34}$ J·s) and ν (the lowercase Greek letter *nu*) is the frequency of the light. Equation (4.1) is known as the *Bohr frequency condition*, after Niels Bohr who proposed it in 1913. It remains the cornerstone of spectroscopy. For emission spectroscopy, the only light that is emitted are those whose frequencies satisfy Eq. (4.1).

For most types of spectroscopy, atoms and molecules already have states available that have different energies, so that absorbing or emitting light can change the state of the atom or molecule. However, in magnetic resonance techniques like nuclear magnetic resonance (NMR) or electron spin resonance (ESR; also called electronic paramagnetic resonance or EPR) spectroscopy, a splitting of energy levels is imposed by a magnetic field. Because changing the strength of the magnetic field changes the amount of splitting of energy levels, there is a relationship between the magnetic field strength H and the frequency of light absorbed or emitted, as given in Eq. (3.4). This is why these methods are referred to as *resonance* spectroscopies.

The most common way to demonstrate what frequencies/wavelengths/energies of light are absorbed by a sample is to plot a graph that shows the intensity of light passing through a sample versus the fre-

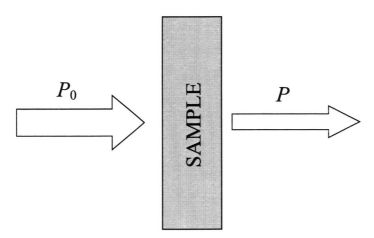

Figure 4.3 The definition of transmittance depends on the incoming light power, P_0, and the transmitted light power, P.

quency/wavelength/energy of the dispersed light. If light of a particular frequency is not absorbed, then the plot hovers around some *baseline*, defined as null signal. If light of a particular frequency is absorbed or emitted, then the plot varies obviously away from the baseline, indicating that light of that frequency satisfies the Bohr frequency condition. Figure 4.2 shows some of these characteristics.

Figure 4.2 shows only one way of displaying a spectrum. There are many ways—indeed, some of the complexity in spectroscopy lies with this fact. Different forms of spectroscopy use different units on the x axes of their spectra and have different ways of indicating a positive signal on the y axes of their spectra. (This is how the axes in spectra are typically assigned.)

As such, the next two sections discuss the common ways of defining the two axes in a spectrum.

4.3 Units of the Y Axis

We will consider the y axis first. There are several common ways of representing the abscissa in a spectrum. Let us begin with a basic one. Dispersed light having original power P_0 at a particular wavelength passes through a sample. If none of the light is absorbed (and we will ignore any reflection or refraction effects), then the transmitted light beam will have power P_0. If some light is absorbed, then the remaining light that makes it through the sample will have some lesser power P. Figure 4.3 illustrates this.

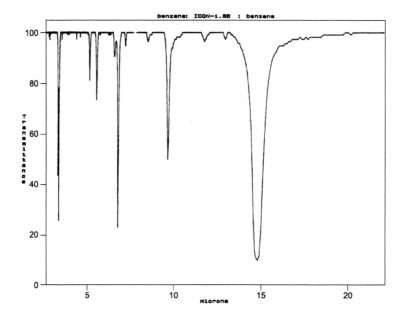

Figure 4.4 A transmittance infrared (vibrational) spectrum of benzene, C_6H_6. The spectrum is plotted versus wavelength of light, in micrometers (microns).

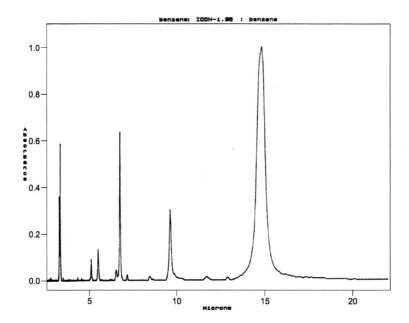

Figure 4.5 The same spectrum as Figure 4.4, but plotted as absorbance versus wavelength of light.

The *transmittance, T,* of the sample is defined as the ratio of the incident power and the resulting power:

$$T = \frac{P}{P_0}.$$ (4.3)

As defined, transmittance is always a numerical value between zero (all light absorbed) and one (no light absorbed). It is common to convert transmittance into *percent transmittance, %T,* by multiplying the original definition by 100%:

$$\%T = \frac{P}{P_0} \times 100\%.$$ (4.4)

Percent transmittance varies between 100% (all light transmitted) and 0% (no light transmitted).

When plotting a spectrum as a transmittance spectrum, 100% is typically at the top of the plot and 0% is at the bottom. Ideally, the baseline of the spectrum is at 100% transmittance, because the baseline is defined as "no signal present." Various factors influence the exact position of the baseline, and it is common to see baselines on transmittance spectra that are not at 100%. Discussion of such factors is outside the scope of this book; details can be found in other texts on spectroscopy and instrumental methods. Figure 4.4 shows an infrared spectrum plotted as percent transmittance. It shows the characteristic "dips" at those wavelengths that are being absorbed by the sample.

Absorbance, A, is defined as the negative base-10 logarithm of the (decimal, not percent) transmittance:

$$A = -\log T = -\log\left(\frac{P}{P_0}\right).$$ (4.5)

Absorbance values range from 0 (no absorbance) to ∞ (all light absorbed). An absorbance spectrum is plotted with the baseline at the bottom of the plot, not the top like in a transmittance spectrum. Rather than placing ∞ at the top of an absorbance spectrum, absorbance values of 1, 2, or 3 typically "top out" the spectrum. Figure 4.5 shows the same spectrum as Figure 4.4, but as an absorbance spectrum.

Absorbance and transmittance are *logarithmically* related, not linearly related. Table 4.2 lists equivalent absorbance and transmittance values.

Table 4.2 Relationship between equivalent absorbance and transmittance values.

Percent of absorbed light	Transmittance	Absorbance
0	1.00	0
90%	0.1	1
99%	0.01	2
99.9%	0.001	3
99.99%	0.0001	4
99.999%	0.00001	5
Etc.		

This table illustrates one problem with absorbance values that have high values: variations in high-absorbance values correspond to a very small difference in the amount of light actually absorbed. Thus, spectroscopists are wary of any spectrum that has particularly high values of absorbance. (It usually means that the sample is too concentrated.)

Absorbance can be useful because it can be shown (see the reference by Ingle and Crouch) that the absorbance value is directly proportional to the concentration of the absorbing species:

$$A = abc, \tag{4.6}$$

where c is the concentration of the absorbing species, b is the sample length, and a is a proportionality constant called the *absorptivity*. If c is molarity units and b is in centimeters, then a is replaced by ε, which is termed the *molar absorptivity*. Equation (4.5) should be applied with care, however. As the previous paragraph mentioned, high values of A should be treated carefully because very small differences in the outgoing light intensity lead to large changes in A when the sample absorbs most of the light. Under such conditions, the direct relationship between A and c in Eq. (4.5) is not followed. Equation (4.5), or its similar form in terms of ε, is called *Beer's law* and was introduced in Chapter 2.

Most absorption spectra are plotted as absorbance or transmittance in the y axis. Emission and Raman spectra are plotted in terms of the number of photons emitted at any particular wavelength, and so they resemble absorbance spectra. Their y-axis unit can be "number of photons emitted" or even some "arbitrary intensity unit" and may not be the same from one spectrum to the next.

Magnetic resonance spectra (NMR and ESR/EPR) are also typically plotted as absorbance spectra, as the magnetic field is varied and resonance conditions are established. While NMR spectra are plotted as straight absorbance spectra, ESR and EPR spectra are plotted as *derivative* spectra. This is because ESR absorptions can be overlapping, and plotting the signal as a derivative makes it easier to identify individual transitions.

4.4 Units of the *X* Axis

Although all spectroscopy is concerned with determining the energy difference between two (or more) states in an atomic or molecular system, most spectroscopic techniques do not use energy units to express those differences. In the SI system of units, energy and energy differences are expressed in units of joules (J), but most spectroscopic techniques do not employ energy units directly when plotting a spectrum.

Instead, we use various related units. Some of these units are directly proportional to energy, and some are inversely proportional to energy. Some are not energy *per se*, but can be related to the energy of the transition, as in magnetic resonance spectroscopy. Novice spectroscopists can be confused about which units are used for which type of spectroscopy.

Most forms of spectroscopy use whichever units yield manageable numbers. For example, a vibrational absorption of CO_2 occurs at 2349 cm^{-1}, which is equivalent to a frequency of 70,420,000,000,000 (7.024×10^{13}) s^{-1}. Given the choice, most spectroscopists would rather use the 2349 than the 70 trillion. But the energy of the vibrational transition is the same in both cases.

The different ways of expressing an energy of transition are related by two simple equations. The Bohr frequency condition, Eq. (4.1), requires that the photon that will be absorbed or emitted by an atomic or molecular system must have the same energy as the difference in energies of the two states involved in the transition. Thus, Eq. (4.1) is a link between energy and frequency, a measurable quantity of the photon. Therefore, some forms of spectroscopy express energy changes in terms of frequency of the light absorbed or emitted.

The second equation relates to light itself, specifically its wave properties. A wave's speed is equal to its wavelength multiplied by its frequency. But Einstein proposed—in fact, his theory of relativity is based on the presumption—that the speed of light *in vacuo* is a universal constant. The speed of light, c, is 2.998×10^8 m/s, so the following equation is also applicable to spectroscopy:

$$c = \lambda v, \tag{4.7}$$

where λ is the wavelength and ν is the frequency. The wavelength has units of meters, which can be converted easily to other units of length; frequency has units of (second)$^{-1}$, or Hertz. Because the speed of light is constant, Eq. (4.6) allows us to convert a given frequency of light into its corresponding (and characteristic) wavelengths, with units we can adjust appropriately.

Wavelength is inversely proportional to energy, however, and some forms of spectroscopy find it convenient to use a unit that is directly proportional to energy. Of course, the reciprocal of the wavelength would be directly proportional to energy, so another unit is introduced, the *wavenumber*:

$$\text{wavenumber} \equiv \tilde{\nu} = \frac{1}{\lambda}. \qquad (4.8)$$

The symbol $\tilde{\nu}$ (read as 'nu-tilde') is commonly used to represent wavenumber. Do not confuse it with ν, the frequency! (It's commonly referred to as a "frequency," but that is another misnomer.) Wavenumber has the SI unit of m^{-1}, but usually more manageable numbers are provided using the non-SI unit cm^{-1} for wavenumber. Wavenumber can be thought of as the number of light waves per meter or per centimeter, depending on the unit used.

Rotational, vibrational, and electronic spectra are usually given in terms of units that take advantage of Eqs. (4.1), (4.6) and (4.7). Magnetic resonance spectra are expressed differently because in most cases a magnetic field is varied to detect absorption of radiation. Thus, transitions in magnetic resonance spectra can be described by the magnetic field strength in units of gauss (G) or, less commonly, tesla (T). There are also ways of expressing the difference in the nuclei's perceived magnetic field in terms of an internal standard, which is how the "parts per million" (ppm) is defined. Interested readers are urged to check the references for additional detail.

The point is that various units are used on the x axis of spectra, but ultimately all of them are related to energy. One of the first tasks when performing any spectroscopic technique is to understand how the ordinate unit relates to ΔE, which is what spectra measure.

Almost all forms of spectroscopy have more than one common x-axis unit. Some of them are related to energy in the same way, some are related differently. For example, compare Figure 4.5, an absorbance spectrum plotted versus wavelength, with Figure 4.6, the same absorbance spectrum but plotted versus wavenumber. Note how the spacing of the absorptions dif-

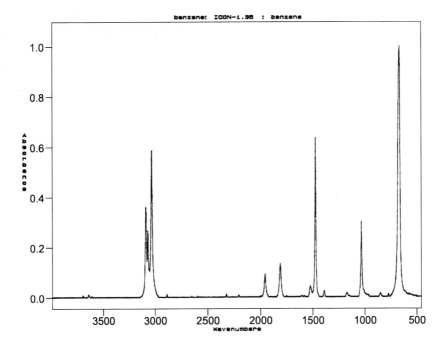

Figure 4.6 The same spectrum as Figs. 4.4 and 4.5, but plotted as absorbance versus wavenumber.

fers depending on the x-axis unit, as well as the widths of the individual signals.

4.5 Typical Examples

Rather than inundate the reader with a flood of spectra to illustrate the possible combinations of x- and y-axis styles, Table 4.3 gives a summary. Notice that most of the variety occurs in the x-axis style; most y axes are plotted as absorbance (for absorption spectra) or some "intensity" of emitted light. With practice, a spectroscopist can tell at a glance which x-axis style represents a direct proportionality to energy and which are inversely proportional to energy.

Table 4.3 Spectrum axis styles for various spectroscopies.

Region	Spectroscopy	X-axis options	Y-axis options
Radio waves	NMR	ppm, Gauss "chemical shift"	absorbance
Microwave	ESR	Gauss	absorbance (derivative)
	(pure) rotational	MHz, GHz, cm^{-1}	absorbance, "intensity"
Infrared	vibrational	μm, cm^{-1}	absorbance transmittance
Visible	electronic	cm^{-1}, nm, Å	absorbance, "intensity"
Ultraviolet	electronic	nm, Å	absorbance, "intensity"
X ray	inner electronic	eV, Å	absorbance
Gamma rays	Mössbauer	cm/s	absorbance

References

1. J. D. Ingle and S. R. Crouch. *Spectrochemical Analysis*, Prentice-Hall, Englewood Cliffs, NJ, 1988.
2. D. A. Skoog, D. M. West, F. J. Holler. *Analytical Chemistry: An Introduction*, Saunders, Philadelphia, 1994.
3. R. S. Drago. *Physical Methods for Chemists* Saunders, Philadelphia, 1992.

Chapter 5
THE SHAPES OF SPECTRAL SIGNALS

5.1 Introduction

One of the powers of spectroscopy is the ability to deduce the quantity of an absorbing material in a sample. As mentioned in the last chapter, this concept is called Beer's law. Here, we will take a closer look at Beer's law, particularly from a historical perspective.

Beer's law relates the absorbance of a signal to the concentration of the absorbing species. This implies that we are interested in the *height* of the line. So aside from the straightforward relationship in Beer's law, what are the factors that contribute to the height of a spectral signal? We will explore some of those factors here.

Also, we should recognize that spectral lines do not appear at certain, exact frequencies. We sometimes speak of them that way, such as "we have an absorption at 687.22 nm," as if the signal were perfectly monochromatic. The reality is, signals are not perfectly monochromatic, and a signal will, upon close inspection, exist over some range. That is, all spectral lines have some *width* to them. We will also consider factors that impose a width to spectral signals.

5.2 The Heights of Lines

In the last chapter, we introduced several versions of Beer's law. The version of Beer's law we will consider here is

$$A = \varepsilon bc,\tag{5.1}$$

where ε is the molar absorptivity of the sample. Remember, of course, that the particular value of the absorbance depends on the wavelength of light sent through the sample. The molar absorptivity ε is dependent on the species absorbing the light, while b and c are experimentally determined parameters. In addition, ε is different at different wavelengths. At some wavelength, ε is large and the sample absorbs light; at other wavelengths, ε is small (or zero) and the sample does not absorb light. In other words, ε is a function of wavelength. Therefore, Eq. (5.1) might be better written as

$$A(\lambda) = \varepsilon(\lambda)bc,\tag{5.2}$$

where the wavelength dependence on the absorbance and the molar absorptivity is now specified. Generally speaking, a plot of $A(\lambda)$ vs. λ is the absorbance spectrum.

The issue of spectral peak intensities focuses on the molar absorptivity $\varepsilon(\lambda)$, and the factors that make it large or small. $\varepsilon(\lambda)$ can be very large, 10^4—10^5 L/cm·mol, for some absorptions, or 0 L/cm·mol for molecules that do not absorb or at wavelengths that are not absorbed by the sample. But, what determines ε?

Well, let us back up a bit. Spectroscopy is a quantum-mechanical phenomenon and quantum mechanics provides the most basic approach to understanding the absorption of light. The quantum mechanical tool of perturbation theory clarifies how a photon affects the state of an atomic or molecular system.

If the initial state of the system before interaction with a photon is given by the wavefunction Ψ_{init} and the final state of the system after interaction with a photon is represented by the wavefunction Ψ_{final}, then quantum mechanics ultimately defines the transition moment (M) as

$$M = \int \Psi^*_{final} \hat{M} \Psi_{init} d\tau, \tag{5.3}$$

where $\hat{M}$ is the appropriate electric dipole operator for the transition. Group theory is useful in understanding Eq. (5.3), because unless the wavefunctions and the operator have the correct symmetries, the value of the integral is exactly 0 and the transition is *forbidden*. If the symmetries are correct (correct symmetries depend on the overall symmetry of the molecule, so it is difficult to be more specific here), then the integral is not necessarily 0 and may range from very small to very large; this is an *allowed* transition.

In the early part of the 20th century, Einstein derived coefficients that relate the probability of an absorption or emission process.[1] For absorption, the *Einstein coefficient of stimulated absorption B* is

$$B = \frac{8\pi^3 M^2}{3h^2 d_{init}}, \tag{5.4}$$

where M is the transition moment, h is Planck's constant, and d_{init} is the degeneracy of the ground state. (Einstein also derived coefficients for spon-

---

[1] Einstein did this in 1917, thus predating quantum mechanics.

taneous emission and stimulated emission; the idea of stimulated emission is crucial to lasers, so in a sense it was Einstein who developed the theory behind lasers.)

It is theoretically useful to treat the electronic or molecular system like a harmonic oscillator, because the wavefunctions for an ideal harmonic oscillator are known, and the expressions for M in Eq. (5.4) can be evaluated. In such a case, it can be shown that

$$B = \frac{e^2\lambda}{4\varepsilon_0 hcm_e},$$ (5.5)

where e is the charge on the electron, λ is the wavelength, ε_0 is the permitivity of free space (necessary to convert the electron charge units to SI units), c the speed of light (do not confuse with concentration!), and m_e is the mass of the electron.

Finally and ultimately, the relationship between the Einstein coefficient as given in Eq. (5.4) and the molar absorptivity ε is

$$\varepsilon(\lambda) = \frac{8\pi^3 n_i \lambda M^2}{2.303 \cdot 3hc^2 d_{init}} \cdot S(\lambda),$$ (5.6)

where n_i is the number of absorbing molecules per cubic centimeter of sample. The $S(\lambda)$ is a line shape function, like a Gaussian or Lorentzian function, that dictates the exact shape of the absorption.

We purposely omitted a few steps leading up to Eq. (5.6). We wanted to find out what absorptivity was made of, and Eq. (5.6) tells us. Many of the variables in Eq. (5.6) are constants, like the speed of light, Planck's constant, π, and so forth. Ultimately, M, as defined by quantum mechanics, will dictate the strength or weakness—that is, the height—of a line in a spectrum.

5.3 Beer's (?) Law

Why is there are a question mark in the section title above? It is because the simple relationship between absorbance, path length, and concentration is sometimes known by different names. Some people refer to it as the Beer–Lambert law, and a very few as the Bouguer–Beer–Lambert law. All three people, apparently independently, developed some of the ideas that are neatly represented in the expression $A = \varepsilon \cdot b \cdot c$.

According to textbooks on chemical history, Pierre Bouguer first noted in 1729 that the amount of light passing through sample decreases with the thickness of the sample. In modern mathematical terms, if we use the variable b to represent the thickness of the sample, Bouguer is saying that

$$A \propto b. \tag{5.7}$$

This was Bouguer's contribution, although it seems clear from the historical record that Bouguer did not propose a mathematical model himself. This seems a very straightforward relationship to us, doesn't it? But to put this in perspective, let us remind ourselves that this point in history, the nature of light was still being debated, the modern atomic theory was almost 100 years away, and Boyle's work on what we now call gas laws was performed only about 70 years earlier.

In the mid-1750s, Lambert rediscovered the relationship between the intensity of the transmitted light and the thickness of the sample. Lambert, however, gave a mathematical relationship to describe the diminution of light intensity. Again in modern mathematical terms, Lambert found that

$$\frac{dI}{I} = -a \cdot dx, \tag{5.8}$$

where a was a constant Lambert called opacity and dx is the infinitesimal distance through the sample. Integrating this expression, converting to base-10 logarithms and grouping constants, and substituting for the definition of absorbance gives the relationship originally announced by Bouguer—i.e., that the absorption of light is related to the distance the light travels through the sample. Although Lambert did recognize that the amount of absorbing "particles" was also a factor in the diminution of light, he apparently failed to model this factor mathematically.

In 1852, August Beer narrowly scooped a French scientist named Bernard in publishing a relationship between concentration and absorption of light. (Indeed, perhaps Bernard's name should be included in the title of the law! It is unfortunate that his contributions have been forgotten.) By passing light through a filter and making it almost monochromatic, Beer noted that the amount of light absorbed was related to the amount of solute contained in various aqueous solutions. In addition, Beer performed studies using sample tubes of different lengths but with the different dilutions of sample, showing that if the amount of sample were the same, the amount of light absorbed is also the same.

However, it seems clear that Beer did *not* derive the expression we call Beer's law. That was demonstrated in 1951 by Pfeiffer and Liebhafsky, when they reproduced some key paragraphs of Beer's original paper. In the first place, it was clear that Beer was referring to *amount of solute*, not concentration. Physical scientists prefer intensive variables (that is, quantities that are independent of amounts, like density), and while concentration is an intensive variable, amount of solute is extensive. In the second place, Beer did propose a formula to model his experiments:

$$\lambda = \mu^D,$$ (5.9)

where Beer labeled μ the absorption coefficient and D the length of the sample, apparently in decimeters. The variable λ was Beer's "relative diminution," akin to the modern measurable transmittance.

Pfeiffer and Liebhafsky did show, however, that Beer's expression is consistent with the modern form of Beer's law. They also suggest that calling this equation "Beer's Law" is a misnomer because Beer did not formulate it *in toto* or in this form, and the references listed below demonstrate that Beer was not the only person to contribute to its development.

Just as we are stuck with the inaccurate name "oxygen" (from the Latin *acid producer*, so named by Lavoisier because he thought it was an essential component of all acids), spectroscopy is saddled with a seemingly inaccurate name for a very simple relationship. However, this very simple relationship seems to have a very complex history.

5.4 The Widths of Lines

In Eq. (5.6), we referred to a lineshape function $S(\lambda)$. We also conceded earlier that spectral signals are not infinitely sharp, but have some finite width. What's the deal with the shapes, or widths, of spectral lines?

In the classroom (usually in physics or physical chemistry courses) when the topic turns to quantum mechanics, many teachers emphasize the idea that energy levels of atoms and molecules are quantized; that is, they have specific values. This means, of course, that changes in energy values also have specific values. This idea leads to the central issue of spectroscopy, the Bohr frequency condition.

The statements mentioned above imply that spectral lines should be very, very sharp. After all, a specific energy change is equivalent to a certain, specific wavelength of light, and so only that wavelength should be represented in a spectrum. Spectra should be composed of almost infinitely sharp lines. But most spectroscopists know that in reality, this never happens. Figure 5.1, for example, shows one peak of the gas-phase HCl rovibrational spectrum. The range of this peak starts at approximately 2910 cm^{-1},

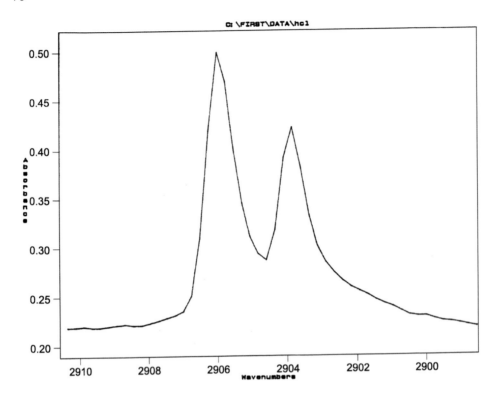

Figure 5.1 An absorption from the rovibrational spectrum of HCl showing a definite width, indicating a range of wavelengths/energies of light being absorbed. The doublet nature of the absorption is due to the natural abundances of ^{35}Cl and ^{37}Cl isotopes.

peaks at 2905.81 and 2903.88 cm^{-1}, and ends at approximately 2900 cm^{-1}; this is definitely not infinitely sharp! And we say that changes in energy are quantized? Sure—but several mechanisms contribute to the broadening of the absorption.

The expansion of a spectral transition from an infinitely sharp line (as might be implied by quantum mechanics) to one with finite width (as found in reality) is called *line broadening*. There are several well-recognized reasons why spectral lines have width. Here, we discuss several of them.

Line shapes are usually approximated by a mathematical function, typically either a Lorentzian function or a Gaussian function. A spectral line whose shape can be described as a Lorentzian function has the general formula

$$F(v) = K\left[\frac{1}{1 + (v - v_0)^2}\right],$$

(5.10)

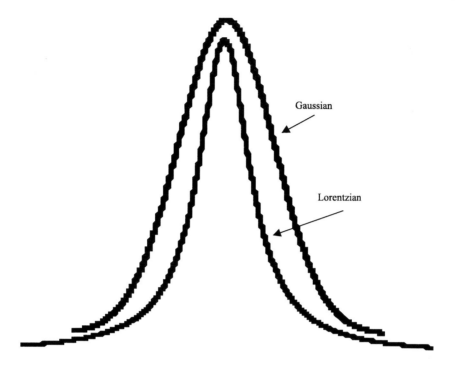

Figure 5.2 Renditions of Gaussian and Lorentzian line shapes. A Voigt profile ranges between both extremes, and most spectroscopic signals lie between a Gaussian and a Lorentzian shape.

where K is some constant, v_0 is the center frequency of the line (and what would be the exact position of the absorption or emission if the line was infinitely narrow), and v is the frequency. A spectral line whose shape can be described as a Gaussian function has the general formula

$$F(v) = Ke^{-k(v-v_0)^2}, \tag{5.11}$$

where K and k are constants. Spectroscopists who study line widths usually use these two functions to "fit" an experimental spectral line. Real spectral lines are usually convolutions of these two types of functions; such convolutions are called *Voigt profiles*. See Ingle and Crouch for additional information. Examples of Lorentzian- and Gaussian-type lineshapes are shown in Figure 5.2.

Line broadening can be either homogeneous or inhomogeneous. Homogeneous line broadening occurs when all atoms or molecules in a sample contribute to the line shape, while inhomogeneous line broadening occurs when different atoms or molecules contribute to different parts of the overall line shape.

Several types of homogeneous line broadening mechanisms exist, but perhaps most fundamental is lifetime broadening. It derives from a perturbation-theory treatment of the interaction of light with quantum states (in other words, spectroscopy), and is reminiscent of the uncertainty principle.[2] Excited states usually persist for a certain period of time before returning to a lower (possibly ground) state. A measure of how long a particular excited state will last is indicated by its lifetime, τ. The uncertainty in the energy of the excited state, ΔE, is related to the lifetime by

$$\tau \cdot \Delta E \approx \frac{\hbar}{2}. \tag{5.12}$$

Spectroscopically, ΔE translates to absorption or emission across a range of frequencies or wavelengths, instead of a certain, specific energy. This imparts width to a spectral line. Since the relationship in Eq. (5.12) is inescapable, the width of a spectral line will not be less than this, and so the line width that occurs because of the relationship in Eq. (5.12) is sometimes referred to as the *natural line width*. Natural line widths can be extremely small and are usually only obtained under highly specialized experimental conditions.

Gas-phase samples are sometimes very prone to broadening. *Collision* or *pressure broadening* is caused by collisions between the gaseous species, which can slightly alter the energies of the ground or excited states. It can be minimized (as its name suggests) if the spectrum of a sample is measured at low pressures. Collision broadening results in a Lorentzian line shape. An effect similar to collision broadening is called *wall broadening* and is caused by the gas-phase sample interacting with its container. Again, however, proper experimental conditions minimize the effects of wall collisions.

Doppler broadening occurs when a gas-phase species moves toward or away from the sources and/or detector. In either case, the apparent energy of the transition changes, widening the spectral line. Because the velocities

---

[2] It does not come from the uncertainty principle directly, although the relationship between energy and time is reminiscent of the uncertainty relationship between position and momentum. The formal mathematical statement of the uncertainty principle is given in terms of the quantum-mechanical operators for the relevant observables. Time does not have a quantum-mechanical operator (although energy does). The relationship in Eq. (5.12) is derived from perturbation theory (see Ch. 6).

of gas-phase species follow a Gaussian distribution, Doppler broadening imposes a Gaussian line shape on the line with the form

$$F(v) = \frac{0.9394}{\Delta v} e^{-2.773(v-v_0)^2/\Delta v^2}, \tag{5.13}$$

where Δv represents the half width at the half maximum of the spectral band. Because different gas-phase species contribute differently to the line broadening, Doppler broadening is an example of an inhomogeneous effect.

Power saturation also affects the overall line shape. It occurs when the incident light beam is so strong that absorption occurs faster than the excited states can decay, and the population of atomic or molecular systems in the ground (or lower-energy) stated is actually depleted. Interested readers can consult molecular spectroscopy texts for a mathematical discussion of this effect. Power saturation can be easily countered by simply attenuating the incident radiation.

Inhomogeneous line broadening can be especially problematic in condensed samples, where interactions between species are impossible to eliminate. For example, inhomogeneous broadening is seen in a molecule that can hydrogen bond, like H_2O. The different orientations of the individual water molecules that are interacting with other water molecules effect the exact vibrational energy of the molecules involved. Since individual molecules interact slightly differently, the cumulative effect is the broadening of the vibrational spectrum. Even samples such as those trapped in rare gas matrices or single crystals are subject to some inhomogeneity that broadens spectral lines.

Finally, instrumental factors also contribute to finite line widths of peaks. Perhaps the easiest to illustrate is for a dispersive spectrometer that uses a slit to let light into a monochromator and out toward a sample. Because the slit has some physical width, a range of light frequencies are being exposed to a sample. As long as at least some light of the right frequency is passing through the sample, some of its power will be attenuated and a signal will be present.

References

1. A. Ihde, *The Development of Modern Chemistry*, Dover Press, New York, 1984.
2. A. Beer, *Ann. Physik Chem.* 86, 78 (1852).
3. F. Bernard, *Ann. Chem. et Phys.* 35, 385 (1852).

4. P. Bouguer, *Essai d'optique tosur la gradation de la lumière*, Paris, 1729.
5. J.D. Graybeal, *Molecular Spectroscopy*, McGraw-Hill, New York, 1988.
6. J.D. Ingle, S.R. Crouch, *Spectrochemical Analysis*, Prentice-Hall, Englewood Cliffs, 1988.
7. H.G. Pfeiffer, H.A. Liebhafsky, *J. Chem. Educ.* 28, 123 (1951).

Chapter 6
QUANTUM MECHANICS AND SPECTROSCOPY

6.1 Introduction

To date, the physical theory that most accurately describes the behavior of the atomic world is quantum mechanics. Since spectroscopy uses light to probe the atomic and molecular world, it should not be much of a surprise to learn that quantum mechanics is used to describe how spectroscopy works.

Nobel Prize-winning physicist Richard Feynmann once said, "If you think you understand quantum mechanics, you don't understand quantum mechanics." This chapter won't bring you to a complete understanding of quantum mechanics, Feynmann's quote notwithstanding. But it will try to point out the important aspects, and hopefully it will at least leave the reader with the understanding that spectroscopy is quantum-mechanically based.

6.2 The Need for Quantum Mechanics

Science's job is to try to understand the natural universe. To do this, scientists try to generate models to describe and predict the universe's behavior. (Not the behavior of the entire universe, but small selected parts of it.) If the model and the universe do not agree, there are two choices: change the model, or change the universe. Since all attempts at changing the universe have failed, our only choice is to change the model.

Since the 1600s, scientific advance has accelerated at least in part because the proposed models of the universe have been increasingly more acceptable. Perhaps the first true "scientific" investigation (by the modern definition) was Robert Boyle's investigation of gases in the 1660s. While there were some noteworthy incorrect or improper models developed (phlogiston and vitalism are two that come to mind immediately, but doubtless there are others), incorrect or improper models were ultimately replaced by more viable models.

One such group of models was what we now call Newton's laws of motion. In just a few statements, verbal or mathematical, Newton's laws accurately describe the motion of matter. Further, these models withstood the test of multiple examinations, and are accepted as proper models of how matter behaves.

To some extent, Newton's laws of motion were so completely accepted that scientists expected all matter to behave that way. The same could be said of Maxwell's laws of electromagnetism, which associated the behavior of electric and magnetic phenomena. As scientists further probed the nature of the universe, they expected that the universe would follow these models.

By the late 1800s and into the early 1900s, the study of nature had progressed to the point that certain phenomena were being studied but could not be explained by the models developed to that point. Ultimately, it was realized that a new model (or models) was necessary.

Briefly, the phenomena were as follows (in no particular order).

The Photoelectric Effect: When light was shone on a metal that was in a vacuum, under certain circumstances electrons would be ejected from the metal. The kinetic energy of the ejected electron was not related to the intensity of the light (as scientists expected it would be), but instead was related to the frequency of the light. Scientists of the 19th century were unable to explain this phenomenon.

Line Spectra of Elements: In the 1860s, it was conclusively demonstrated that elements had their own unique emission spectra, consisting of sharp lines of light of certain wavelengths. The emission spectrum of hydrogen was particularly simple: in the visible light region of the spectrum, it was composed of four lines having certain wavelengths. In fact, Balmer was able to show that the wavelengths fit the following equation:

$$\frac{1}{\lambda} = R\left(\frac{1}{4} - \frac{1}{n^2}\right), \tag{6.1}$$

where R was a constant and n equaled 3, 4, 5, and 6 for the particular lines. After similar groups of lines were found in the emission spectrum of hydrogen in other regions of the spectrum (like the infrared and ultraviolet), Rydberg found that all of these groups of lines could be predicted by the equation

$$\frac{1}{\lambda} = R_H\left(\frac{1}{n_1^2} - \frac{1}{n_2^2}\right), \tag{6.2}$$

where n_1 and n_2 were integers. (R is now called the Rydberg constant in his honor.) This was all well and good, but the question remained: *why* was this so? Other elements had their own characteristic emission spectra, but no simple mathematical equation was found to predict the wavelengths of those lines of light.

Atomic Structure: With the discovery of electrons and protons (and, ultimately, neutrons), the atomic theory was modified somewhat: atoms *are* divisible, and they are composed of subatomic particles. Rutherford's experiments in the 1910s supported the idea of a nuclear structure of atoms, with the heavy protons (and neutrons) in a central nucleus, and the electrons "orbiting" the nucleus at some distance. The problem with that came from Maxwell's electromagnetic theory. Accelerating charges emitted energy, and if electrons are in curved orbits around the nucleus, they then should be constantly emitting energy, eventually losing all their energy and crashing down into the nucleus. Thus, by Maxwell's laws, a nuclear atom should be inherently unstable. But matter *was* stable (as far as we could tell, except for radioactive matter, but we won't consider that here), so either the nuclear model was wrong or Maxwell's laws were wrong—or some other model was needed.

The Nature of Blackbody Radiation: A blackbody is a perfect absorber of light. Since absorption and emission are opposite processes, it follows that a blackbody is also a perfect emitter of light. This suggests that a heated blackbody should emit light of all wavelengths equally. However, experimental measurements showed that not all wavelengths are emitted equally. Figure 1.7 in Chapter 1 shows the relative intensities of different wavelengths of light emitted by a blackbody. Although there were some attempts to explain this behavior using the current understanding (most notably, by what we now call the Rayleigh-Jeans law), no model was able to accurately predict the behavior of blackbody radiation.

Low-temperature Heat Capacities: By the beginning of the twentieth century, temperatures approaching absolute zero were finally attainable in the laboratory. Scientists determining the properties of matter at such low temperatures discovered that the heat capacities of matter decreased as the temperature itself decreased. Einstein developed a useful model for this, and Debye developed a similar, more applicable model, but the question remained: why did matter behave this way?

Collectively, these issues that could not be explained by the theories of the time meant that new ideas and theories would be necessary.

6.3 Planck's Theory and Einstein's Application

The study of blackbody (or cavity) radiation made several advances before Planck performed his work. For example, in 1859, Gustav Kirchhoff (the same Kirchhoff who helped develop the spectroscope) showed that the distribution of intensities of blackbody radiation was (a) independent of the material used to construct the blackbody, and (b) dependent only on temperature and wavelength. Thus, there was some distribution function I that was universal for all blackbody radiation. In 1879, Josef Stefan—apparently based on only a few experimental measurements—proposed that the total power per unit area emitted as light varied with the fourth power of temperature. This suggestion was derived theoretically by Ludwig Boltzmann in 1884, so the expression

$$P = \sigma T^4, \tag{6.3}$$

is called the *Stefan-Boltzmann law*. The value of σ, the Stefan-Boltzmann constant, is 5.670×10^{-8} W/m$^2 \cdot$ K^4. Finally, in 1893, Wilhelm Wien noted that the maximum wavelength in the plot of blackbody radiation, λ_{max}, times the absolute temperature gave a constant:

$$\lambda_{max} \cdot T = \text{constant} = 2.898 \text{ mm} \cdot \text{K} . \tag{6.4}$$

Equation (6.4) is called the *Wien displacement law*. As T increases, λ_{max} shifts to smaller and smaller values. Finally, in Chapter 1 we mentioned how the Rayleigh-Jeans law attempted to model blackbody radiation, but only a partial fit was found for the long-wavelength side of blackbody radiation; an *ultraviolet catastrophe* was actually predicted because of the rapid increase in predicted intensity for small-wavelength light.

Max Planck was a thermodynamicist by training, and he decided to approach the issue of blackbody radiation from a thermodynamic point of view. In particular, Planck presumed that if some energy were introduced into a cavity at a particular temperature (i.e. a blackbody), that energy would ultimately rearrange into some equilibrium distribution of energy values. From that perspective, and with a little mathematical "trick," Planck produced a formula in 1900 that predicted the intensity of blackbody radiation over its whole range.

The trick is an interesting one, actually. Planck treated light as interacting with tiny electric oscillations in matter, recognizing the electromagnetic nature of light. Classically, the energies of oscillators are related to the amplitudes of the oscillation. But Planck took the mathematical assumption that the energy of an oscillator was proportional to its *frequency*, not its amplitude. For reasons that are debated today, Planck used the letter h to represent the proportionality constant:

$$E \propto \nu$$
$$E = h\nu. \tag{6.5}$$

With that little twist, Planck derived what we now call the Planck radiation law, which in terms of Kirchhoff's distribution function I is

$$I(\lambda, T) = \frac{2\pi hc^2}{\lambda^5}\left(\frac{1}{e^{\frac{hc}{\lambda kT}} - 1}\right). \tag{6.6}$$

(In fact, it can be shown that the Rayleigh-Jeans law is a Taylor-series approximation of Planck's distribution law—an approximation that ultimately did not work very well.) Using this expression, Planck was able to derive both the Wien displacement law and the Stefan-Boltzmann law.

Although successful in predicting the shape of the blackbody intensity curve, most scientists—even Planck—did not assign any physical significance to the relationship in terms of light itself. But in 1905, Einstein did. Using Planck's ideas, Einstein was able to explain the nature of the photoelectric effect *if one assumed that light itself had an amount of energy given by the equation E = hν.* Thus, Einstein extended the relationship between energy and frequency to light, not just to the electric oscillators in the matter composing the blackbody. Essentially, light was acting like a "particle" of energy. It took a while to be accepted—it was, after all, contrary to light's well-established wave nature—but in 1921, Einstein was awarded the Nobel Physics Prize, *not* for his work in relativity but for his work on the photoelectric effect.

What Planck's and Einstein's work established is that energy comes in discrete quantities, rather than any possible quantity. Consider the following analogy of a car in motion. It has kinetic energy equal to $\frac{1}{2}mv^2$, where m is the mass of the car and v is its velocity. The kinetic energy can have *any* value from zero (corresponding to a zero velocity) and higher. At the level of individual light waves, however, light can have only certain *specific* values of energy, depending on what its wavelength is. That is, the energy of light is *quantized*. Furthermore, remember the problem of atomic emission spectra, which are composed of lines of light of certain particular frequencies—and therefore, of certain particular energies. It seems obvious that the electrons in atoms themselves have quantized energies, if their differences are restricted to certain values. But, that is discussed in the next section.

Note, however, that Planck's and Einstein's work on the *quantum theory of light* has explained two phenomena that previous science could not: the behavior of blackbody radiation and the photoelectric effect. The dichotomy of perspectives in science before and after Planck's ideas are so stark that 1900 is considered the dividing point in time between Classical Physics (pre-1900) and Modern Physics (post-1900).

6.4 Bohr's Model

Still unexplained were atomic spectra, although some progress was made with the introduction of quantized energy. In 1913, after meeting with famous physicists like Rutherford—who was promoting the concept of a nuclear atom—Danish physicist Niels Bohr made a few assumptions and, on the basis of those assumptions, was able to algebraically derive the expression for the spectrum of the hydrogen atom as proposed by Rydberg. That is, Bohr was able to model the spectrum of the hydrogen atom.

Bohr's assumptions were the following:

- Electrons "orbited" the nucleus in a circular orbit, with the coulombic attraction between proton and neutron balanced by the centrifugal force of the orbiting electron.

- Electrons maintain a constant energy as they orbit. Thus, despite the fact that electrons are charged particles, Bohr is assuming that Maxwell's laws simply do not apply.

- Electrons can change orbits by absorbing or emitting light, but only if the energy of the light equals the difference in the energies of the orbits. Mathematically, this is given by $\Delta E = h\nu$. This is called the *Bohr frequency condition*.

- The angular momentum, mvr, of the electron in its orbit is quantized, and is limited to integral values of $h/2\pi$. Mathematically, this is given as $mvr = \nu h/2\pi$, where n is the integral value.

With these assumptions, Bohr was able to derive the following equation for the difference in energies of the hydrogen atom:

$$\frac{1}{\lambda} = \frac{m_e e^4}{8\varepsilon_0^2 h^3 c}\left(\frac{1}{n_i^2 - n_f^2}\right). \qquad (6.7)$$

This equation has the same form as the Rydberg equation, with $R = m_e e^4 / 8\varepsilon_0^2 h^3 c$. The variables n_i and n_f are integral values attached to the angular momentum (see Bohr's assumptions, above). Thus, Bohr was able to predict the spectrum of the hydrogen atom, if the assumptions he made were correct. Notice that one of the assumptions is that the electrons have a constant energy in their orbits; thus, the energy of electrons in the hydrogen atom were *quantized*. Bohr was also able to derive an expression for the energy of the electrons in their orbits, and the expressions consisted of constants like m_e (the mass of the electron), h, e (the charge on the electron)—and n. The parameter n is called a *quantum number*, because it alone determines the overall energy of the electron. (The other quantities are constants.) The radius of the first orbit is a convenient atomic scaling factor and is called the *first Bohr radius*, a_0:

$$a_0 = \frac{\varepsilon_0 h^2}{\pi m_e e^2} = 0.529 \text{ Å}. \tag{6.8}$$

A radius of 0.529 Å implies a diameter of about 1 Å, and experimental evidence of the time was suggesting that hydrogen atoms were about that size.

Though an obvious breakthrough, Bohr's theory was limited to hydrogen and any other single-electron systems (like He^+, Li^{2+}, etc.). Other advances in the next few years suggested that Bohr's assumptions were naïve. Two advances in particular were the uncertainty principle and the concept of de Broglie waves. The Uncertainty Principle is the idea that certain related variables have limits in the level of accuracy with which they could be determined. One set of related variables is position and linear momentum. If the limit of accuracy in position is represented as Δx and the limit of accuracy in momentum is Δp_x, then the Uncertainty Principle requires that

$$\Delta x \cdot \Delta p_x \geq h/4\pi. \tag{6.9}$$

At normal scales, these accuracy limits are unnoticeable. But on the atomic scale, they can be relatively substantial. Bohr's theory is suggesting that we can determine the quantized momentum and the exact distance of the electron from the nucleus, in apparent contradiction of the Uncertainty Principle!

Louis de Broglie pointed out another problem. His basic thesis was, if light (a wave) can have particle properties (in terms of energy), why couldn't tiny particles (like electrons) have wave properties? In 1923, de Broglie deduced the expression

$$\lambda = \frac{h}{p}, \tag{6.10}$$

where p is the momentum of a particle of matter and λ is its *de Broglie wavelength*.

For normal matter under normal conditions, the de Broglie wavelength is undetectable. But for an electron in Bohr's first orbit (i.e. $n = 1$), λ is predicted to be 3.3 Å—three times the diameter of a hydrogen atom! This suggests that the wave behavior of tiny particles like electrons should not be ignored in developing a model of atomic behavior, as Bohr's theory of hydrogen ignores it. Thus, Bohr's model—while useful and a major step forward in terms of its quantized angular momentum—is limited.

6.5 Quantum Mechanics

In 1925–26, Werner Heisenberg and Erwin Schrödinger published two different but mathematically equivalent theories on the behavior of atomic-level systems that incorporated the Uncertainty Principle and de Broglie's theory. Heisenberg's theory is based on matrix algebra, while Schrödinger's theory is based on a second-order differential equation. Schrödinger's mathematics are more approachable for most scientists, and is the more common formalism. Here, we will focus exclusively on Schrödinger's version of *quantum mechanics*.

Schrödinger proposed that the state of a system was described by a wavefunction Ψ, which contains all information about the system. Possible values of observables (like position, momentum, energy, etc.) could be obtained as *eigenvalues* of an *eigenvalue equation*. An eigenvalue equation is a mathematical equation that combines an operator, $\hat{O}$, with a (wave) function, Ψ, to yield some constant, K, multiplying the original function:

$$\hat{O}\Psi = K \cdot \Psi. \tag{6.11}$$

You should resist that temptation to say that $\hat{O}$ equals K, because the operator $\hat{O}$ is oftentimes a differential expression (i.e., it contains a derivative) or a function itself. Schrödinger proposed that every observable had a corresponding operator; for example, the linear momentum operator $\hat{p}$ in the x dimension was given by the expression

$$\hat{p} = -i\hbar \frac{d}{dx},\qquad(6.12)$$

where $\hbar = h/2\pi$ and i is the square root of -1.

Schrödinger said that the wavefunction for the state of a system could be anything, but had to satisfy the eigenvalue equation for energy, given by

$$i\hbar \frac{d\Psi}{dt} = \hat{H}\Psi.\qquad(6.13)$$

Because there is a time derivative of the wavefunction in this equation, Eq. (6.13) is called the *time-dependent Schrödinger equation*. $\hat{H}$ is the operator that is connected to the total energy of the system. If the time variable on the wavefunction were separable from the spatial variables (x, y, and z in three-dimensional space), then Ψ could be written as

$$\Psi = (e^{-iEt/\hbar}) \cdot \psi(x, y, z).\qquad(6.14)$$

The variable E is the total energy of the system. The function $\psi(x, y, z)$ is the time-independent wavefunction and will be represented simply as ψ. Under these conditions, the wavefunction ψ itself must satisfy the following three-dimensional *time-independent Schrödinger equation*:

$$\left[-\frac{\hbar^2}{2m}\left(\frac{\partial^2}{\partial x^2} + \frac{\partial^2}{\partial y^2} + \frac{\partial^2}{\partial z^2}\right) + \hat{V}\right]\psi = E \cdot \psi\qquad(6.15)$$

Equation (6.15) is also an eigenvalue equation, with the eigenvalue being E, the energy of the system. The operator on the left side of the equation is called the Hamiltonian operator, and is given the symbol $\hat{H}$—the same one found in the time-dependent Schrödinger equation. [Equation (6.15) is also a second-order differential equation.] $\hat{V}$ is the operator for the potential energy of the system. Equation (6.15) is sometimes abbreviated

$$\hat{H}\Psi = E\Psi.$$

Though perhaps Eq. (6.15) looks complicated, it is simply a second-order differential equation, and finding wavefunctions Ψ that satisfy this equation are the purview of a branch of mathematics. Schrödinger (and other developers of quantum mechanics) was well-schooled in mathematics, and was able to apply Eq. (6.15) to several ideal physical systems. He provided solutions to Eq. (6.15) (i.e., he found wavefunctions ψ that satisfied the differential equation) for a particle-in-a-box, an ideal harmonic oscillator, and a particle of mass moving about a center in two and three dimensions (i.e., moving in a circle and on the surface of a sphere). In all cases, Schrödinger found that by imposing physical restrictions on the mathematical solutions for Ψ, the total energy of the system could not be any value, but rather was quantized to certain specific values. In all cases, integers called quantum numbers naturally occur in the wavefunctions.

In addition (and most relevant to us as spectroscopists), Schrödinger attacked the hydrogen atom. According to Bohr, the hydrogen atom could be treated as a particle (the electron) moving in a circle about a center (the nucleus). This corresponded to Schrödinger's two-dimensional circular movement—but the quantized energies for motion in a circle did not agree with the spectrum of the hydrogen atom. Schrödinger defined the hydrogen atom differently: not only could the electron appear at any angle around the nucleus—that is, it existed spherically around the nucleus, not just circularly—but it could also exist at *any* possible distance, rather than being confined to particular orbits.

With these assumptions, Schrödinger used the time-independent Schrödinger equation to predict *the same spectrum of the hydrogen atom that Bohr's theory did*. Thus, Schrödinger's version of quantum mechanics was able to model a physically real system. If a model agrees with reality, then there has to be some truth to the model, right?

There was some concerns, however, that the concept of a wavefunction violated the uncertainty principle. If quantum mechanics assigns a wavefunction to an electron in hydrogen, isn't it violating the Uncertainty Principle just as Bohr's 'orbits' violate it? Doesn't finding a wavefunction imply that we know exactly where the electron is? And is the electron a particle or a wave?

The Copenhagen Interpretation was formulated by Bohr and Max Born in the late 1920s to deal with these difficulties. (It was called that because much of the work was done at Bohr's Institute for Atomic Studies, which was located in Copenhagen, Denmark.) In particular, Born proposed that rather than giving the exact path of the electron around the nucleus, the square of the modulus of the wavefunction, $|\Psi|^2$, was proportional to the probability that the electron existed at any particular point in space. Thus, quantum mechanics is not stating precisely where the electron is, but only its probability of being somewhere.

This interpretation, and other conclusions based in it, was subject to a lot of criticism and resistance from many scientists, including prominent ones like Schrödinger, Einstein, and de Broglie. It reduces the atomic world to one of statistical probabilities instead of events specifically determined by previous events. Einstein's objection is summed up in his famous quotation, "God does not play dice." Unfortunately for Einstein, decades of theory and experiment (including some ground-breaking experiments in the 1980s) supports the Copenhagen Interpretation as the best current interpretation of a quantum-mechanical wavefunction, so it is proper to say that God *does* play dice—He just doesn't play with loaded dice.

Since the mid-1920s, quantum mechanics has been developed further and been applied to matter at the atomic, molecular, and bulk scales; and it has generally been shown to be an amazingly successful model of the behavior of matter. Quantum theories of nuclear behavior have also been successfully developed. In fact, one of the first achievements of quantum mechanics was to explain radioactivity (one of the topics not specifically discussed in Section 6.2). Tunneling, the phenomenon by which particles can penetrate an energy barrier even if they do not have enough energy to "climb over" the barrier, was not only invoked to explain alpha decay of radioactive nuclei, but was also harnessed to construct scanning tunneling microscopes in the 1980s. The theory of relativity was incorporated into quantum mechanics by Paul A. M. Dirac in 1928. In so doing, Dirac was able to predict the existence of antimatter, which in turn was conclusively identified by Carl Anderson in 1932 with the discovery of the positron. Relativity in quantum mechanics also predicted the concept of electron spin, upon which a form of resonance spectroscopy is based. (All subatomic particles have spin.) In 1927, Walter Heitler and Fritz London devised the first quantum-mechanical treatment of the hydrogen molecule. Quantum mechanics can be invoked to understand the low-temperature behavior of the heat capacities of solids. The behavior of electrical conductors, nonconductors, semiconductors, and even superconductors can be explained using quantum mechanics.

Quantum mechanics sounds like the ultimate model of the universe, doesn't it? Well, it's not. Quantum mechanics does have certain limitations. The most important one for our purposes here relates to the form of the Schrödinger equation in Eq. (6.15). (This is the *non*relativistic form of the Schrödinger equation.) Recall that it is a second-order differential equation in three dimensions. Recall, too, that there is a branch of mathematics that focuses on finding solutions to such differential equations and eigenvalue equations. There are certain tactics to apply in finding such solutions, one of which is to invoke *separation of variables*. Separation of variables is the assumption that a function of several variables can be separated into the product of simpler functions, *each* of which is a function of a *single* variable:

$$F(x, y, z) = f(x) \cdot f(y) \cdot f(z). \qquad (6.16)$$

Effectively, separation of variables simplifies the solution of the eigenvalue equation by breaking it down into solvable equations of a single variable. The solution of the entire problem is then the combination of the smaller solutions. The complete wavefunctions for the electron in the hydrogen atom can be determined this way, for example. We say that we can solve the Schrödinger equation *analytically* for the hydrogen atom; we can determine specific functions which, when substituted into Eq. (6.15), are exact mathematical solutions to the differential equation.

One problem with the application of quantum mechanics to matter is that atoms or molecules with more than one electron have wavefunctions that *are not separable*. We cannot write a wavefunction for, say, helium that has a part of the wavefunction completely determined by one electron and another part of the wavefunction completely determined by the second electron. The reason is that the electrons interact with each other, since they are both negatively charged. Without having a system whose mathematical description is separable, we would have to solve the differential equation for all variables simultaneously. To date, that has not been accomplished, and there is reason to believe that it is analytically impossible. Therefore, the conclusion is that we cannot find analytic solutions to the Schrödinger equation for as small an atom as helium.

Does this mean that quantum mechanics is useless? No—several examples of the applicability of quantum mechanics were listed above. What it does mean is that we need other tools to apply quantum mechanics to other systems. There are two main tools that we can use to apply quantum mechanics and approach systems *numerically*, rather than *analytically*. Ultimately, we can apply quantum mechanics numerically to any degree of accuracy we want (or have the time for), and it works. Therefore, we find that analytical solutions to the Schrödinger equation are not needed.

The two tools are called variation method and perturbation theory. The variation method is based on the variation theorem, which states that any guess, called a trial function, for the true wavefunction will always give a value for energy that is higher than the true energy. Operationally, adjustable parameters are placed in the trial wavefunctions, the Schrödinger equation is used to determine the energies of these wavefunctions, and values of the parameters are determined that yield the lowest possible energies for that trial function.

One advantage of variation theory is that the trial wavefunctions can be *any* function, as long as it is physically relevant to the system. Also, any number of variables can be included in the trial wavefunction. A problem with variation theory is that the trial wavefunction focuses on energy as an observable and not any other observable, and that the greater the number of variables, the greater the mathematical complexity. But there is one saving grace: most of the mathematics can be performed by computer. These days, computers are fast enough and computer programs are available to perform the mathematics needed for variation theory. Therefore, variation theory is a large part of the computational quantum mechanics being performed by scientists around the world.

Perturbation theory is the other tool used to apply quantum mechanics to atoms and molecules, and it has special relevance to spectroscopy. We will treat it in the next section. Understand, however, that with these two tools, quantum mechanics provides a model for understanding atomic and molecular spectra. Quantum mechanics thus provides an understanding of all outstanding problems that classical mechanics could not, and because of that should be treated as the superior theory.

6.6 Perturbation Theory

Perturbation theory assumes that a system can be approximated as a known, solvable system, and that any differences between the system of interest and the known system is a small, additive perturbation that can be calculated separately and added on. Perturbation theory assumes that the Hamiltonian operator for a real system can be written as

$$\hat{H}_{system} \approx \hat{H}_{ideal} + \hat{H}_{perturb} \tag{6.17}$$

where $\hat{H}_{system}$ is the Hamiltonian operator of the system of interest that is being approximated, $\hat{H}_{ideal}$ is the Hamiltonian operator of an ideal system, and $\hat{H}_{perturb}$ represents the small, additive perturbation. If we assume that the wavefunction Ψ of the real, nonideal system is similar to the wavefunction of the ideal system Ψ_{ideal}, then one can say that, *approximately,*

$$\hat{H}_{system} \Psi_{ideal} \approx E_{system} \Psi_{ideal} \,. \tag{6.18}$$

Applying the appropriate mathematics (which will not be presented here), one can ultimately get a relationship between the energy of the real system, E_{system}, and the energy of the ideal system, E_{ideal}, which should be known. The relationship is

$$E_{system} = E_{ideal} + \int \Psi_{ideal}^{*} \hat{H}_{perturb} \Psi_{ideal} \,, \tag{6.19}$$

where the superscript "*" on the first wavefunction implies that we have taken the complex conjugate of the wavefunction (i.e. substituted $-i$ for i wherever the square root of -1 appears in the wavefunction, if at all). The integral in Eq. (6.19) can either be solved analytically by considering a table of integrals, or can be solved numerically. Thus, we can use perturbation theory to define a real system in terms of an ideal system and determine the energy of the real system. For example, the two electrons in a helium atom can be defined as two hydrogen-electron systems (the ideal systems whose wavefunctions and energies are known) plus an electrical repulsion between the two negatively charged electrons (the $\hat{H}_{\text{perturb}}$). Applying this to helium allows us to determine an energy for helium that is ~5% within the experimental value. Not bad!

Perturbation theory is useful because *any number* of additive perturbations can be added to the ideal Hamiltonian operator. However, perturbation theory does not share variation theory's guarantee that the lower the energy means the better the wavefunction. Still, many scientists are using perturbation-theory-based models to understand atomic and molecular systems.

6.7 Application to Spectroscopy

In spectroscopy, a system is exposed to an oscillating electromagnetic field that has some variation in time. We can apply perturbation theory to the system, with the understanding that the perturbation is the oscillating electromagnetic field. This field varies in time, so what we are considering is called *time-dependent* perturbation theory.

The Hamiltonian operator for the system is designated $\hat{H}^0$. The operator for the oscillating perturbation (i.e., the electric field of the light) is $\hat{H}'(t)$, and is given by the expression

$$\hat{H}'(t) = H'\cos 2\pi vt = H'\frac{(e^{i2\pi vt} + e^{-i2\pi vt})}{2} \tag{6.20}$$

In Eq. (6.20), v is the frequency of the light, and t is time. The second form of the operator comes from Euler's relation between cosine (and sine) and imaginary exponentials. If this operator were used as the perturbation and perturbation theory was applied, we would ultimately find (and the details are omitted here) that the probability, $P(t)$, of the system being in an excited state would be given by the expression

$$P(t) = \frac{4 \cdot \left| \int \Psi_{\text{final}}^* \hat{H}'(t)\Psi_{\text{initial}} \right|^2}{(2\pi v_\Delta - 2\pi v)^2} \cdot \sin^2\left[\frac{1}{2}(2\pi v_\Delta - 2\pi v) \cdot t\right], \tag{6.21}$$

where Ψ_{final} and $\Psi_{initial}$ are the final- and initial-state wavefunctions, respectively; v is the frequency of the light involved; and v_Δ is the difference in energies of the final and initial states, expressed in frequency units.

Equation (6.21) looks complex, but the important parts are the sine term and the denominator that includes the frequency terms. The remaining terms (the 4 and the term that includes the integral) are constants. Thus, we can simplify Eq. (6.21) by writing it as a proportionality:

$$P(t) \propto \frac{\sin^2\left[\frac{1}{2}(2\pi v_\Delta - 2\pi v) \cdot t\right]}{(2\pi v_\Delta - 2\pi v)^2}. \qquad (6.22)$$

Eq. (6.22) is similar to an expression called the *sinc function*:

$$\text{sinc}(x) = \frac{\sin \pi x}{\pi x} \qquad (6.23)$$

[More precisely, Eq. (6.22) is related to the *square* of Eq. (6.23).] The sinc function has a very distinctive plot: it approaches 1, its maximum value, when the variable πx approaches zero, but quickly falls off to zero (in an undulating fashion, as expected for a sine function) as πx moves away from zero.

How is this relevant for us? Well, in Eq. (6.22) above, the "variable" is $(2\pi v_\Delta - 2\pi v)$. As that value approaches 0, the sinc function in Eq. (6.22) approaches 1 (its maximum value), but as the quantity $(2\pi v_\Delta - 2\pi v)$ moves away from zero, the value of the sinc function falls off rapidly toward zero—and so does the probability $P(t)$ of the system being in an excited state! Another way to say this is, as the quantity $(2\pi v_\Delta - 2\pi v)$ approaches zero, there is an increased probability that the system will absorb light and move to an excited state, but as the quantity $(2\pi v_\Delta - 2\pi v)$ deviates from zero, there is a lesser probability that the system will absorb light and change to an excited state.

And when does $(2\pi v_\Delta - 2\pi v)$ approach zero? When $2\pi v_\Delta = 2\pi v$, or more simply, when

$$v_\Delta = v. \qquad (6.24)$$

If we multiply both sides by Planck's constant h, we get

$$h v_\Delta = h v; \qquad (6.25)$$

and now we can say that the probability of a system absorbing light and changing its wavefunction is highest when the energy of the light, hv, equals the energy difference between the two states, hv_{Δ}.

Thus, the tools of quantum mechanics can be used to explain spectroscopy. Interested readers are urged to consult the references for more detailed mathematics in the topic.

References

1. P. Atkins, R. S. Friedman, *Molecular Quantum Mechanics*, 3rd edition, Oxford University Press (1997).
2. D. Ball, *Physical Chemistry*, Brooks-Cole Publishing Company, Pacific Grove, CA (2003)
3. J. Gribbin, *In Search of Schrödinger's Cat*, Bantam Books, New York (1984).

Chapter 7
SELECTION RULES

7.1 Introduction

By now, it should be understood that a system will only absorb (or emit) light if the energy of the light equals the energy difference between the two states involved in a spectroscopic transition. But, there is no guarantee that the transition will occur if this circumstance is met. That is, the Bohr frequency condition is a necessary but not sufficient condition for a transition to occur.

Additional criteria for whether an absorption or emission occurs are called *selection rules*. Selection rules can be grouped into two types, either quantum-mechanical or descriptive (sometimes called "gross selection rules"). That is, some selection rules can be explained using quantum mechanics and wavefunctions and operators and quantum numbers. Other selection rules can be explained by describing what the atom or molecule is doing.

If a transition is favored by a selection rule, we say that the transition is *allowed*. If a transition does not follow a selection rule, we say that the transition is *forbidden*. However, since most selection rules are formulated with the assumption of ideality and real systems are not ideal, some forbidden transitions may actually occur. This can understandably confuse the interpretation of a spectrum!

In this chapter, we will look at where these selection rules come from (with the short answer being, of course, "quantum mechanics"!). First we will consider the more formal mathematical perspective, then we will review a simpler approach based on changes in quantum numbers.

7.2 "Dipole Moment" Selection Rules

Spectroscopists use selection rules to keep track of what transitions are allowed or forbidden. Although most selection rules are derived by assuming quantum-mechanically ideal systems, in reality most selection rules are not followed to the letter. But selection rules are still helpful in understanding spectra.

Some selection rules are given in terms of allowed changes in quantum numbers. These rules are fairly specific: $\Delta v = \pm 1$ for vibrations, or $\Delta J = \pm 1$ for rotations. We will consider these in the next section. But some selection rules are a bit more general. Two of them, in particular, are based on the dipole moment of a molecule:

- In order to have a pure rotational spectrum, a molecule must have a permanent dipole moment. (Example: HCl does, but H_2 and Cl_2 do not.)

- In order for a vibration to absorb light and appear in a spectrum, there must be a change in the dipole moment of the molecule associated with that vibration. (Example: All three vibrations of H_2O appear in water's vibrational spectrum, but not all vibrations of CO_2 do.)

Where do these general selection rules come from?

First, recall that a dipole moment μ is defined as a charge separation, e, over some distance $\mathbf{r}$:

$$\mu = e\mathbf{r} \tag{7.1}$$

Here, both μ and $\mathbf{r}$ are in bold, indicating that they are vectors. As mentioned in Chapter 5, quantum-mechanically, a spectroscopic transition is *electric-dipole allowed* if the following integral is nonzero:

$$M = \int \Psi_1^* \hat{\mu} \Psi_2 d\tau, \tag{7.2}$$

where Ψ_1 and Ψ_2 are the upper and lower wavefunction of the (electronic, vibrational or rotational) state, $d\tau$ is the general three-dimensional-space infinitesimal required by the integration, and $\hat{\mu}$ is the *dipole moment operator*, which is given by an expression analogous to Eq. (7.1):

$$\hat{\mu} = e\hat{r}. \tag{7.3}$$

The operator $\hat{r}$ is the position operator. The quantity M in Eq. (7.2) is called the *transition moment*. M may be exactly zero, in which case the transition between Ψ_1 and Ψ_2 is forbidden. If it is not zero, the transition is allowed.

The dipole moment operator $\hat{\mu}$ is simply multiplicative. There is no derivative-taking or any other change in the wavefunctions. However, $\hat{\mu}$ does contain the variable r, and so do Ψ_1 and Ψ_2. Therefore, the integral in Eq. (7.2) may be zero or nonzero, depending on the properties of the product $\Psi_1 \cdot r \cdot \Psi_2$. The evaluation of this integral, and the quantum-mechanical conditions under which it must be exactly zero or might be nonzero, ultimately leads to specific selection rules in terms of changes in quantum numbers (i.e., $\Delta l = \pm 1$, $\Delta v = \pm 1$, $\Delta J = 0, \pm 1$, etc.). However, we are looking for selection rules in terms of the dipole moment, not the quantum numbers.

First, let us address rotational transitions. How do we get the statement that for a pure rotational spectrum, the molecule must have a nonzero dipole moment? Consider a molecule rotating in three-dimensional space, where the rotations are separated into components in the three dimensions x, y, and z. The dipole moment vector μ can also be separated:

$$\mu = \mu_x + \mu_y + \mu_z. \tag{7.4}$$

At this point we will introduce the unit vectors **i**, **j**, and **k**. The vector **i** points in the positive x direction and has magnitude of 1. The unit vectors **j** and **k** serve the same purpose for the y and z dimensions, respectively. Thus, we can rewrite Eq. (7.4) in terms of the magnitude of each component multiplied by the proper unit vector:

$$\mathbf{m} = \mu_x \mathbf{i} + \mu_y \mathbf{j} + \mu_z \mathbf{k}. \tag{7.5}$$

The magnitudes μ_x, μ_y, and μ_z are now scalar, not vector, quantities.

Quantum mechanics supplies an understanding of three-dimensional rotations for ideal systems, and that understanding is applied to rotational spectroscopy. But to simplify this understanding, rotational motion is considered in terms of spherical polar coordinates r, θ, and ϕ. Simple geometry can relate x, y, and z with r, θ, and ϕ. Table 7.1 lists these relationships, while Figure 7.1 illustrates the two coordinate schemes.

The dipole moment μ can be written in terms of spherical polar coordinates instead of Cartesian coordinates. If μ_0 is used to represent the (scalar) magnitude of the dipole moment, then Eq. (7.5) becomes

Table 7.1 Cartesian-to-spherical polar coordinates (and vice versa).

$x = r \cdot \sin\theta \cdot \cos\phi$	$r = \sqrt{x^2 + y^2 + z^2}$
$y = r \cdot \sin\theta \cdot \sin\phi$	$\theta = \cos^{-1}\left(\dfrac{z}{\sqrt{x^2 + y^2 + z^2}}\right)$
$z = r \cdot \cos\theta$	$\phi = \tan^{-1}\left(\dfrac{y}{x}\right)$

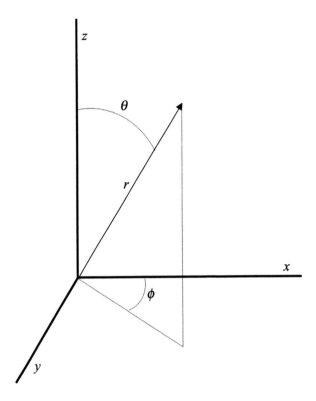

Figure 7.1 The relationship between Cartesian and spherical polar coordinates. See Table 7.1 for the mathematical relationships.

$$\mu = \mu_0(\sin\theta \cdot \cos\phi \cdot \mathbf{i} + \sin\theta \cdot \sin\phi \cdot \mathbf{j} + \cos\theta \cdot \mathbf{k}). \tag{7.6}$$

The operator $\hat{\mu}$ has a similar form. Therefore, the expression for the transition moment M becomes

$$M = \int \Psi_1^* \times \mu_0(\sin\theta\cos\phi \cdot \mathbf{i} + \sin\theta\sin\phi \cdot \mathbf{j} + \cos\theta \cdot \mathbf{k})\Psi_2 d\tau. \tag{7.7}$$

The magnitude of the dipole moment μ_0 is a constant and so can be brought outside of the integral. We get

$$M = \mu_0 \cdot \int \Psi_1^* \times (\sin\theta\cos\phi \cdot \mathbf{i} + \sin\theta\sin\phi \cdot \mathbf{j} + \cos\theta \cdot \mathbf{k})\Psi_2 d\tau. \tag{7.8}$$

What Eq. (7.8) says is that if a molecule does not have a *permanent* dipole, then μ_0 is exactly zero and therefore M is exactly zero! Therefore, a mole-

cule must have a nonzero dipole in order to have a pure rotational spectrum.

What about vibrations? In a vibration, the positions of the atoms are
oscillating about some equilibrium position. Since the dipole of a molecule
depends on the positions of the atoms, the exact value of μ is a function of
the atomic positions; that is,

$$\mu = f(\mathbf{r}), \tag{7.9}$$

where $f(\mathbf{r})$ is some as-yet unknown function of $\mathbf{r}$. Since $\mathbf{r}$ is changing in a
vibration, we need an expression for how μ varies with $\mathbf{r}$, too. Any well-
behaved function can be expressed in terms of a Taylor series. In the case of
the dipole moment, we have

$$\mu = \mu(\mathbf{r}_e) + \left.\frac{\partial \mu}{\partial \mathbf{r}}\right|_{\mathbf{r}_e} \cdot \Delta \mathbf{r} + \frac{1}{2}\left.\frac{\partial^2 \mu}{\partial \mathbf{r}^2}\right|_{\mathbf{r}_e} \cdot (\Delta \mathbf{r})^2 + \ldots \tag{7.10}$$

In Eq. (7.10), each derivative is evaluated at the position $\mathbf{r} = \mathbf{r}_e$, the equilibrium position of the vibration. If we approximate the dipole moment μ
using only the first two terms of Eq. (7.10), the transition moment M for a
vibrational transition is

$$M = \int \Psi_1^* \left[\mu(\mathbf{r}_e) + \left.\frac{\partial \mu}{\partial \mathbf{r}}\right|_{\mathbf{r}_e} \cdot \Delta \mathbf{r} \right] \Psi_2 \, d\tau.$$

We can expand into two integrals and, recognizing that $\mu(\mathbf{r}_e)$ and $\left.\frac{\partial \mu}{\partial \mathbf{r}}\right|_{\mathbf{r}_e}$ are
constants, remove them from the integrals to get

$$M = \mu(\mathbf{r}_e) \cdot \int \Psi_1^* \Psi_2 \, d\tau + \left.\frac{\partial \mu}{\partial \mathbf{r}}\right|_{\mathbf{r}_e} \int \Psi_1^* \cdot \Delta \mathbf{r} \cdot \Psi_2 \, d\tau. \tag{7.11}$$

Because different vibrational wavefunctions are orthogonal, the first integral in Eq. (7.11) is zero. (This is one useful mathematical property of wavefunctions that we did not cover in Chapter 6.) The expression for M
becomes

$$M = \left.\frac{\partial \mu}{\partial \mathbf{r}}\right|_{\mathbf{r}_e} \int \Psi_1^* \cdot \Delta \mathbf{r} \cdot \Psi_2 \, d\tau. \tag{7.12}$$

Now consider this equation, like we did equation Eq. (7.8) for rotations. The partial derivative $\partial\mu/\partial\mathbf{r}|_{\mathrm{r_e}}$ is evaluated at the equilibrium position $\mathbf{r_e}$. It has a particular value, some number. However, if the dipole moment μ does not change, the value of that derivative at the equilibrium distance is exactly zero, and the transition moment will be exactly zero. Therefore, in order for a vibrational transition to be allowed, there must be a changing dipole moment associated with that vibration.

Of course, almost all of these equations are meant for ideal systems, and in reality transitions that are strictly forbidden will sometimes occur. There are also selection rules based on magnetic moments and quadrupole moments, too. But at least for most straightforward applications in spectroscopy, these "dipole moment" selection rules are the ones to know.

7.3 Symmetry Arguments for M

The discussion in the previous section rationalized why particular molecular motions (rotations and vibrations) absorb or emit light, but for any form of spectroscopy, the fundamental relationship that indicates whether a transition is allowed or forbidden is Eq. (7.2):

$$M = \int \Psi_1^* \hat{\mu} \Psi_2 \mathrm{d}\tau,$$

A transition moment M can be defined for any type of state-to-state transition of an atom or molecule, so this expression is central to understanding selection rules.

Consider the expression, however. It is an integral of the product of three functions. (One of the terms is an operator, but that operator is a multiplicative one, and does not require derivatives or any other change of Ψ_2.) The product of three functions is simply another function, and an integral is simply an area under a curve. So the transition moment is simply an area under the curve of the function that results from the product of three functions. An area is a number; therefore, the transition moment is simply some *number*. (The limits on the integral go from $+\infty$ to $-\infty$, but because of the properties of acceptable wavefunctions, the area under the curve is *never* infinite.)

It may seem that all we need to do is evaluate the integral, but that is easier said than done. Remember, we do not have mathematical solutions for wavefunctions for anything other than the hydrogen atom, so all wavefunctions for other systems are approximations, if we have them at all. However, thanks to a very useful property, we do not have to solve the integral because we can show that the integral for a particular transition is exactly zero (and therefore the transition is forbidden). That useful property is *symmetry*.

Symmetry is the concept that defines the spatial and dimensional similarity of an object. We speak of things having "higher" or "lower" symmetry by simply recognizing that some objects are more dimensionally similar than others. Consider a rectangle and a square, for example. A square is more symmetric than a rectangle because all of its four sides are the same length and all of its angles are 90 deg. A rectangle normally has different-length sides perpendicular to each other (although we recognize that a square is actually a special kind of rectangle). A cubical object is more symmetric than a solid object that is not cubical, unless it is perfectly spherical. Molecules have symmetry, too. The benzene molecule (C_6H_6) is a hexagonal molecule, while methane (CH_4) has the a tetrahedral shape.

The formal mathematical study of symmetry is called group theory. Group theory is a very useful tool in spectroscopy, because *wavefunctions themselves also have symmetry*. Thus, we can use the ideas of group theory to understand wavefunctions and, in the case of transition moments, products of wavefunctions.

Let us use Figure 7.2 to illustrate what we mean. Figure 7.2(a) shows a semicircular function. The function has the interesting symmetry property of having the same value at $-x$ as it has at x. [Mathematically, $f(x) = f(-x)$.] Such functions are called *even functions*. On the other hand, the wavy function in Figure 7.2(b) has the symmetry property of having the opposite value at $-x$ as it has at x. [Mathematically, $f(x) = -f(-x)$.] Such functions are called *odd functions*.

If we were to determine the area under the curve shown in Figure 7.2(a), we would get some nonzero value. However, if we were to determine the area under the curve shown in Figure 7.2(b), we would get exactly zero. That is because the area in the positive section on the left side of the function is numerically canceled by the area in the negative section on the right side. Generally speaking, the net area under the curve of an odd function will always be *exactly zero* (as long as the interval is symmetric).

Group theory helps us determine whether wavefunctions are odd or even—and we are able to determine this even without knowing the exact form of the wavefunction. The product of odd and even functions follows the same rules as products of positive and negative numbers:

$$\text{odd} \times \text{odd} = \text{even}$$
$$\text{even} \times \text{even} = \text{even}$$
$$\text{odd} \times \text{even} = \text{odd.}$$

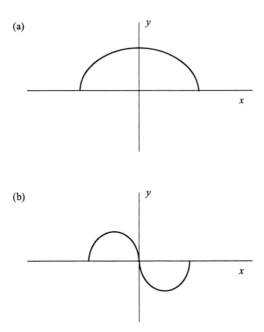

Figure 7.2 Examples of even (a) and odd (b) functions. The integral of the function in (a) is some nonzero value that represents the area under the curve. However, the integral of the function in (b) is exactly zero, because the "positive" area on the left side is cancelled by the "negative" area on the right side. The integral of any odd function over a symmetric interval is exactly zero. This concept helps us recognize that some integrals are exactly zero without actually integrating.

So, after determining whether the wavefunctions involved and the operator are odd or even functions, we use the above rules to determine if the three-part product $\Psi_1^* \hat{O} \Psi_2$ is, overall, odd or even. If it is odd, then the integral in the transition moment is exactly zero. The transition moment integral for many such combinations in atomic and molecular systems turn out to be exactly zero. Only even combinations can have a nonzero value for the integral—and at that, there's no guarantee how large or small the integral might be, but it is no longer constrained to being exactly zero.

Actually, it gets more specific than this (although it gets difficult to describe without reference to group-theoretical language). Every kind of physical symmetry has something called a "totally symmetric representation" within that symmetry. The product of the two wavefunctions and the operator need not just be even in order to have a nonzero integral value. If the overall representation of the three-part product is *not* the all-symmetric representation, then the value of the integral *must* be zero. So it gets more detailed than "odd vs. even." Interested readers are urged to consult texts on group theory and its application to spectroscopy to learn these details. Many selection rules can be determined from group theoretical principles.

7.4 Summary of Selection Rules

The two previous sections give some physical and mathematical rationale for some selection rules. Here, we will present and discuss a summary of the selection rules for various forms of spectroscopy. These selection rules are ultimately based on the physical or mathematical understanding discussed above. Rather than derive each selection rule independently, we will simply state them and leave the search for specific details to the interested reader.

The reader should be advised that this section considers only the selection rules that relate to the electric dipole operator in the perturbation-theory approach to spectroscopy. Electric quadrupole and magnetic dipole operators can also be applied using perturbation theory, and they lead to different selection rules. However, such spectra are typically much less intense and are likely to be less commonly seen in a straightforward spectroscopy experiment.

7.4.1 Electronic spectroscopy

The wavefunctions of the hydrogen atom can be described by four quantum numbers: the principal quantum number n, the angular momentum quantum number l, the z-component of the angular momentum quantum number m_l, and the z-component of the spin angular momentum quantum number m_s. (There is also the spin angular momentum quantum number s, but for all electrons, $s = \frac{1}{2}$.) These four quantum numbers arise from the mathematical solution of the Schrödinger equation, Eq. (6.15), for the hydrogen atom system. When a hydrogen atom changes state by emitting or absorbing a photon, the wavefunction changes and, thus, goes from one set of quantum numbers to another. That is, there is a *change* in one or more quantum numbers. For allowed transitions, the following are the allowed changes in the various quantum numbers:

$$\Delta n = \text{anything}$$
$$\Delta l = +1 \text{ or } -1$$
$$\Delta m_l = 0 \text{ or } +1 \text{ or } -1$$
$$\Delta m_s = 0.$$

There are similar selection rules for multielectron atoms, but the quantum numbers are defined a little differently. For small atoms ($Z < 20$ or so), we treat the electrons as if their orbital angular momenta combine, or *couple*, to produce an overall orbital angular momentum whose values are given by a quantum number L. Similarly, the spin angular momenta couple to produce an overall spin angular momentum whose values are given by a quantum number S. Finally, we recognize that the overall orbital angular momentum and the overall spin angular momentum can also couple, and the resulting total electronic angular momentum is represented by the quantum number J, which has a z-component represented by M_J. For atoms, the selection rules are

$$\Delta L = 0 \text{ or } +1 \text{ or } -1$$
$$\Delta S = 0$$
$$\Delta J = 0 \text{ or } +1 \text{ or } -1$$
$$\Delta M_J = 0 \text{ or } +1 \text{ or } -1,$$

except that $\Delta L = 0$ is forbidden if the initial value of L is 0 and $\Delta J = 0$ is not allowed if the initial value of J is 0.

For larger atoms, we gain a better understanding of electronic behavior if we assume the orbital angular momentum and the spin angular momentum of each *individual* electron couple to make a total electronic angular momentum represented by the quantum number j. However, even for large atoms, it is common for the individual electronic states to be labeled according to the L and S quantum numbers, so the above selection rules are still useful for large atoms as well.

Electronic spectra of diatomic molecules follow similar selection rules, except the relevant quantum numbers are labeled Λ, Σ, and Ω. These represent the magnitude of the orbital electronic momentum along the molecular axis, the projection of the spin angular momentum along the molecular axis, and the vector combination of the first two components, respectively. For diatomic molecules, the selection rules are

$$\Delta\Sigma = 0$$
$$\Delta\Lambda = 0 \text{ or } +1 \text{ or } -1$$
$$\Delta\Omega = 0 \text{ or } +1 \text{ or } -1.$$

Selection rules for electronic transitions in molecules are based on group theory and depend on the symmetry of the molecule. The group-theoretical basis of selection rules was introduced briefly in the previous section.

7.4.2 Pure rotational and vibrational spectroscopy

The rotations of molecules can be treated quantum-mechanically by assuming that the molecule acts as a rigid rotor and can thus be treated as a particle moving on a spherical surface. This is one of the ideal systems that quantum mechanics can treat analytically. In doing so, we derive rotational states that depend on a rotational quantum number J. There is also the z-component of the rotation, indexed by the quantum number M_J. (It is important to not confuse these labels with the labels used to describe the total electronic angular momentum states, although they do come from a similar quantum-mechanical treatment.) However, molecules can have up to three unique axes of rotation, depending on the symmetry of the molecule. Tetrahedrally shaped methane, CH_4, has all-equivalent axes of rotation, so its rotational behavior can be described with a single rotational quantum number. Molecules like methane are referred to as *spherical tops*. The rotations of the bent water molecule, H_2O, have three different rotational axes with different rotational behavior; molecules of this type are called *asymmetric tops*. The pyramidal-shaped ammonia molecule, NH_3, needs only two unique axes to be defined. These molecules are called *symmetric tops*, and there are two distinctions among symmetric tops. *Prolate* symmetric tops have their two larger rotational moments of inertia equal, while *oblate* symmetric tops have their two smaller rotational moments of inertia equal. Symmetric tops also have quantized angular momentum about a molecular axis, and the quantum number K is used to represent the quantized angular momentum about that axis.

Recall from Section 7.2 that there are dipole-moment-related selection rules for rotations. Molecules must have a permanent dipole moment to show a pure rotational spectrum. Because of their symmetry, spherical tops will never have a dipole moment and therefore will not show a pure rotational spectrum.

Prolate and oblate symmetric tops may have a permanent dipole moment, so may show a pure rotational spectrum. When they do, we can describe them in terms of selection rules for changes in J, M_J, and K:

$$\Delta J = 0 \text{ or } +1 \text{ or } -1$$
$$\Delta M_J = 0 \text{ or } +1 \text{ or } -1$$
$$\Delta K = 0.$$

Again, there is the exception that $\Delta J = 0$ is not allowed if the initial value of J is 0. If the rotational spectrum is measured using Raman scattering, we have the rule that $\Delta J = +2$ or -2.

For asymmetric tops, J and M_J are useful quantum numbers, so the first two expressions are good guides as selection rules. However, K is no longer a useful quantum number, and selection rules get more dependent on group theory. We will not consider such cases here.

The vibrations of polyatomic molecules can look complex, but the vibrations of a molecule with N atoms can always be broken down into $3N - 6$ *normal modes of vibration*. These normal modes, collectively, describe all possible vibrations of the molecule. Each vibration can be treated independently as if it were an ideal harmonic oscillator. Quantum mechanics gives us analytic solutions for the ideal harmonic oscillator system, and in doing so introduces a vibrational quantum number v for each vibration.

The quantum-mechanical selection rule for vibrations is straightforward:

$$\Delta v = +1 \text{ or } -1.$$

Mathematically, this selection rule derives from the Taylor-series expansion of μ as a function of $\mathbf{r}$, Eq. (7.10). Recall that we truncated our expansion after only two terms. The selection rule above is based on the second term in Eq. (7.11), which leads to a transition-moment integral having the expression $\Psi_1^* \Delta \mathbf{r} \Psi_2$ in it. However, if we take the Taylor-series expansion to the next term, we would get a transition-moment integral that has the expression $\Psi_1^* (\Delta \mathbf{r})^2 \Psi_2$ in it. This term would ultimately yield some additional allowed transitions that follow the selection rule

$$\Delta v = +2 \text{ or } -2.$$

These transitions, called *overtone transitions*, are typically less intense than those that obey $\Delta v = +1$ or -1. Higher-magnitude changes in Δv are formally allowed by additional terms in the Taylor-series expansion of μ, but are correspondingly less intense in most circumstances. These selection rules hold if you are measuring the vibrational spectrum using Raman scattering as well.

7.4.3 Magnetic resonance spectroscopy

In magnetic resonance spectroscopy, we are taking advantage of the fact that subatomic particles (either electrons or nuclei) have a spin I (or S for the electron). As is true for any spin angular momentum, there are $2I + 1$ (or $2S + 1$) possible orientations of the z-component of the spin angular momentum; this z-component is labeled m_I (or m_S). Normally, the energy of the system is independent of the z-component's quantum number. We say that these wavefunctions are *degenerate*. But in the presence of a magnetic field, the different orientations of the z-component of the spin—that is, the differing values of m_I or m_S—have different energies. Energy in the form of photons can be absorbed and the system can experience a transition from one spin orientation to another spin orientation. That is, there is a change in the m_I (m_S) quantum number. The selection rules for magnetic resonance spectroscopy are

$$\Delta m_I = +1 \text{ or } -1$$
$$\Delta m_S = +1 \text{ or } -1.$$

Despite the simple selection rules, NMR and ESR spectra can be very complex, because the exact resonance condition depends on the bonding environment of the particle involved.

7.4.4 Violations, mixing types of motions

As useful as selection rules are, they are not followed absolutely. For example, vibrational spectra routinely measures *combination bands*, which are absorptions in which $\Delta E = h(\Sigma v_i)$, where an absorption can be assigned to a sum or difference of two or more vibrations of a molecule. Forbidden transitions are commonly observed in the electronic spectra of transition metal ions, and such transitions are actually responsible for the variety of colors seen in transition metal compounds. Formally forbidden transitions are responsible for phosphorescence, a long-lived emission process that is long-lived *because* the transition is actually a forbidden one.

There are other reasons that spectra do not always follow the rules. One is that the Born-Oppenheimer approximation, mentioned in Chapter 6, really is an approximation, and many times we actually probe a transition that is a combination of several molecular processes. Or, one molecular process (like a vibration) might increase the probability that another process will absorb light even though the selection rules don't formally allow it. Thus, we can have *vibronic* spectra, in which electronic transitions are induced by the interaction of vibrational states with electronic states.

Another common circumstance is to have one molecular process superimposed on another molecular process. For example, electronic spectra of molecules commonly show a pattern that is caused by the vibrations of the molecules. Another common occurrence is the superimposition of individual rotational transitions on a single vibrational absorption of a molecule. Figure 7.3 shows an example of such a *rovibrational spectrum* for gas-phase CO_2. Although CO_2 has no dipole moment and thus does not exhibit a *pure*

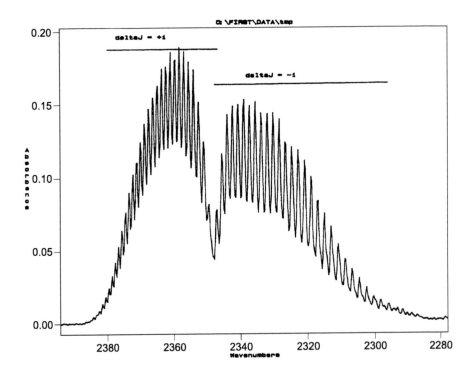

Figure 7.3 A rovibrational spectrum of the asymmetric stretching motion of gas-phase CO_2. This spectrum corresponds to a $\Delta n = +1$ and $\Delta J = +1$ or -1 simultaneously. The regions where ΔJ is $+1$ and where ΔJ is -1 are labeled.

rotational spectrum, this particular vibration distorts the molecule enough to give it a short-term dipole moment. Under those conditions, we see the equivalent of a pure rotational spectrum superimposed on the vibrational transition.

References

1. F. A. Cotton, *Chemical Applications of Group Theory*, 3rd ed., Wiley, New York, 1990.
2. G. Herzberg, *Molecular Spectra and Molecular Structure, I. Spectra of Diatomic Molecules*, Van Nostrand, New York, 1950.
3. G. Herzberg, *Molecular Spectra and Molecular Structure, II. Infrared and Raman Spectra of Polyatomic Molecules*, Van Nostrand, New York, 1945.
4. G. Herzberg, *Molecular Spectra and Molecular Structure, III. Electronic Spectra of Polyatomic Molecules*, Van Nostrand, New York, 1967.
5. J. L. McHale, *Molecular Spectroscopy*, Prentice-Hall, Upper Saddle River, NJ, 1999.

Chapter 8
RESOLUTION AND NOISE

8.1 Introduction

There are several issues that communicate how "good" a spectrum is. In this chapter, we will focus on two of them: resolution and noise. Both are factors that can affect the quality of a spectrum. Keep in mind that not all of the issues discussed here are applicable to all types of spectroscopy. A good spectroscopist should recognize which issues will impact the particular spectrum being measured.

8.2 Resolution in Dispersive Spectrometers

A spectrum, recall, is usually some graphical representation of electromagnetic radiation absorbed or emitted versus the energy of that radiation. One concern of anyone measuring a spectrum should be, how close can two different spectral transitions be and still be differentiated as different signals? This question relates to the idea of *resolution*.

The American Society for Testing and Materials (ASTM) defines *spectral resolution* as

> the ratio $\lambda/\Delta\lambda$, where λ is the wavelength of radiant energy being examined and $\Delta\lambda$ is the spectral bandwidth expressed in wavelength units; or, alternatively, the ratio $v/\Delta v$, where v is the wavenumber of the radiant energy being examined and Δv is the spectral bandwidth expressed in wavenumber units.

That is, the resolution is some unitless numerical value that depends on the wavelength or frequency (i.e., wavenumber) of the light and the spectral bandwidth of the spectrometer's monochromator. For dispersive spectrometers, the spectral bandwidth is the wavelength (or frequency) interval that is coming out of the exit slit or its equivalent. Spectroscopists recognize that no monochromator will pass a single frequency at a time; rather, a range of frequencies always comes out together.

This definition of resolution depends on the spectral bandwidth, which is itself determined by the dispersive ability of the monochromator and the size of the instrument's exit slit. Since the exit slit's width can be controlled by the experimenter or the spectrometer (i.e. automatically), the dispersive ability of the monochromator ultimately determines the resolution.

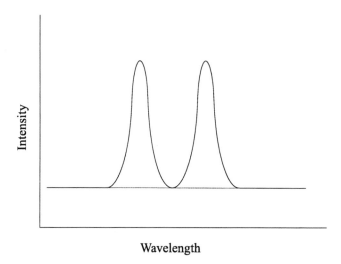

Wavelength

Figure 8.1 The baseline criterion for resolution requires that the spectrum reach the baseline between the two peaks. In this case, the peaks are resolved; if they were any closer (i.e. overlapping), they would not meet this criterion for resolution.

The ability of a monochromator's prism or grating to disperse, or separate, individual wavelengths of light is given by the ratio $\partial x/\partial \lambda$, which relates how far you have to travel in some x dimension to get a specific change in the wavelength λ of light. Typically, the distance traveled is at the exit slit, moving from one side of the slit to the other. This ratio is called the *linear dispersion*. More commonly used is the *reciprocal* of the linear dispersion, $\partial \lambda/\partial x$, which indicates how much the wavelength changes with distance. For example, if a monochromator has a reciprocal linear dispersion of 0.75 nm (of λ) per mm (of x), then a slit that is 2 mm wide will pass a range of wavelengths, $\Delta\lambda$, of 1.50 nm. The spectral bandwidth can therefore be taken as 1.50 nm. (The term *spectral bandpass* is also used, but the ASTM definitions use band*width* as the standard terminology.)

The bandwidth calculated above suggests that resolution changes with wavelength. For a bandwidth of 1.50 nm, a peak at 700 nm would have a resolution of 467, while a peak at 300 nm would have a lower resolution of 200. It is indeed the case. The reciprocal linear dispersion does not change much with wavelength for grating monochromators (although for prism monochromators the change in dispersion can be large).

While there is a specific formula for resolution, to many spectroscopists the working definition of "resolution" is a little different. Any two spectral wavelengths are *resolved* if they can be absolutely differentiated from each other. Resolution is not much of an issue when identifying the positions of two spectral features that are widely separated from each other, although it

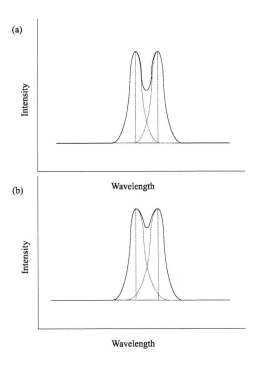

Figure 8.2 The Rayleigh criterion for resolution is less stringent than the baseline crite-rion, and probably used more frequently by spectroscopists. It requires that λ_{max}, the wavelength maximum for the adjacent peaks, be over the baseline and not over any part of the adjacent peak. In (a), the two peaks are considered resolved by this criterion, but not the two peaks in (b). Note that in neither case is the baseline crite-rion for resolution satisfied.

may affect their bandwidth. Resolution is much more an issue with spec-tral features that are close together.

Figure 8.1 shows the *baseline criterion* for resolution. Two absorptions are considered resolved if the light signal reaches the spectrum's baseline between the two peaks. It can be shown (see Ingle and Crouch) that the spectral bandwidth must be less than or equal to one-half of the wave-length difference between the two peaks in order for them to be resolved. Since an increasingly smaller bandwidth means an increasingly higher value for resolution (since $\Delta\lambda$ is in the denominator of the definition for res-olution), we can say that a *higher* resolution allows us to distinguish between *smaller* differences in wavelengths of peaks in a spectrum.

Very close spectral features can overlap, thanks to their intrinsic or experimental linewidths. The *Rayleigh criterion* for resolution requires that if two peaks are overlapping, each peak maximum must be where the other peak has already reached baseline. Figure 8.2 shows examples of peaks that are and are not resolved based on this less-stringent criterion. Most spectroscopists probably use the Rayleigh criterion as the working definition of resolution.

As indicated above, only a few instrumental factors have a large effect on resolution in dispersive instruments. Certainly the exit slit width is one easily controlled factor, although the size of the slit must be selected with other issues in mind, like signal-to-noise ratios and Fraunhofer diffraction effects. Perhaps the most important factor is the dispersive element and the size of the spectrometer. A prism or grating with a higher angular dispersion—which is different from linear dispersion—will separate individual wavelengths better, ultimately providing better resolution. Alternately, a spectrometer can be made longer to allow the light to disperse more before hitting the exit slit (with a decrease in throughput, however). Better dispersing elements seem to be the obviously favored approach!

8.3 Resolution in Fourier Transform Spectrometers

Many of the specifics in the previous section are only applicable to dispersive spectrometers. One of the more common spectrometers is the Fourier transform infrared spectrometer, and resolution issues are a little different for FT machines. Here, we will explore those differences.

In a dispersive spectrometer, both the wavelength and spectral bandwidth are determined by the physical characteristics and settings of the spectrometer. The mirrors and gratings, the entrance and exit slit widths, even the detector system can contribute to λ and $\Delta\lambda$. Hence, they all contribute to the resolution of the resulting spectrum.

Fourier transform (FT) spectrometers are different in that all wavelengths of light pass through the optics simultaneously. One advantage of this is that the detector sees a rather bright light, which is more easily detected than the dimmer signal made by light that has been dispersed by a monochromator. This is known as *Jacquinot's advantage*. But, if all wavelengths are observed simultaneously, does this not imply a large $\Delta\lambda$ and therefore a very low resolution? (Recall that the mathematical definition of resolution is $\lambda/\Delta\lambda$.)

Actually, resolution for FT spectrometers is considered a little differently.

Recall that FT spectrometers measure a spectrum differently from a dispersive instrument. The crucial part of the FT optics is an interferometer, which splits the incoming light beam into two parts. One part travels down a fixed path and is reflected back to the beamsplitter. The other part travels down a path that has a moving mirror. This part of the light therefore experiences a changing path length as it is reflected back to the beamsplitter. The two beams, recombined, pass through the sample and to a detector. Figure 3.5 (back in Chapter 3) shows a simple schematic of an interferometer.

Because all wavelengths of light pass through the interferometer at the same time, massive destructive interference occurs when the split light beams are combined—except when the two light paths equal each other. (Remember, one mirror is moving back and forth.) Then, massive constructive interference occurs. A plot of the recombined light's intensity versus the *difference* in the two pathlengths is called an *interferogram*. This difference, called the *optical path difference*, is designated by the lowercase Greek letter delta, δ. A typical interferogram is shown in Figure 3.6, also back in Chapter 3. The sudden intensity at zero path difference (or ZPD) is called the *centerburst*. (Note that δ is not the amount that the mirror moves, because the light beam must travel back *and* forth before it recombines. The mirror displacement, Δ, is therefore half the value of δ.)

Lord Rayleigh is credited for recognizing that an interferogram (a plot of intensity versus position) is mathematically related to a spectrum (a plot of intensity versus wavelength) by an operation called the *Fourier transform*. However, it was not until the mid-1960s that efficient computer algorithms were developed to make FT instruments practical.

So how is resolution determined for FT instruments?

Perhaps the first point to make is that resolution is usually phrased differently in FT spectra, despite the ASTM definition quoted at the beginning of this chapter. According to that definition, resolution is defined as $\lambda/\Delta\lambda$ or $v/\Delta v$, which implies that a better resolution is expressed as a *higher* number. However, it is prevalent in FT spectroscopy to express resolution as the energy difference (in cm^{-1}) between two closest signals that can be differentiated. By that convention, a *lower* numerical value is associated with a *better* (or "higher") resolution. For example, a 32-cm^{-1}-resolution spectrum can differentiate two absorptions that are separated by 32 cm^{-1}, while a 0.5-cm^{-1}-resolution spectrum can differentiate two absorptions that are separated by only 0.5 cm^{-1}. Thus, the second spectrum has a higher resolution because it can better differentiate closer signals.

To explain how resolution is determined in FT instruments, here we will follow the tactic taken by Smith; interested readers might want to consider Griffiths and de Haseth's book for a more rigorous treatment. Consider two peaks (absorptions, transmissions, or other type of positive signal) that are close to each other. Will they be resolved—using Rayleigh's criterion, at the very least—by the spectrometer? First of all, if we are presuming that the spectrum consists of only two infinitely sharp lines, then the interferogram should be a combination of two sine waves. The two individual sine waves might look something like Figure 8.3. Notice how closely spaced the two sine waves are.

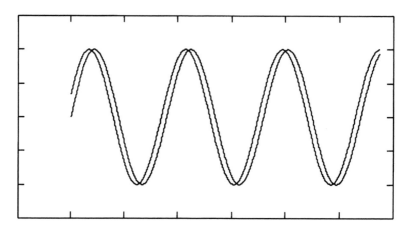

Figure 8.3 Two signals that are 1.0 cm^{-1} apart will contribute sine waves to the interferogram that are very close to each other. In order to resolve the two separate signals, the interferogram must be determined to a certain level of resolution itself.

An FT spectrometer measures a signal that is a combination, or sum, of all of the sine waves. However, it does not record the signal as a continuous waveform; rather, it digitizes the signal into individual, discrete points. (Ultimately, a spectrum is stored as a series of (x,y) points.) In order to resolve the *combined* waveform into its components, the moving mirror must move far enough so that enough interferogram points are digitized to be able to calculate a unique Fourier transformed spectrum. The smaller the mirror displacement Δ, the less number of points that are used to define the interferogram. This means that there is less detail in the digitized interferogram, so the resulting spectrum has a lower resolution. Resolution is therefore inversely proportional to Δ, the mirror displacement:

$$\text{resolution} \propto \frac{1}{\Delta}. \tag{8.1}$$

We can also say that the resolution is also inversely proportional to δ, the optical path difference:

$$\text{resolution} \propto \frac{1}{\delta}. \tag{8.2}$$

Is there a more specific relationship? Consider that our two waves of different frequencies combine to give regularly spaced nodes of complete destructive interference, as well as regularly spaced maxima of constructive interference. This pattern of regular maxima and minima is termed a *beat*. (The same thing happens in music, when musicians play two tones that are slightly out of tune.) The beat has its own frequency, which is determined by the frequency difference of the two component waves:

$$v_{beat} = \Delta v = v_1 - v_2. \tag{8.3}$$

(In this example, the beat frequency would actually have wavenumber units, and so would represent the number of beat "waves" per meter or centimeter.) The beat also has a wavelength that is the reciprocal of its wavenumber value:

$$\lambda_{beat} = \frac{1}{v_{beat}}. \tag{8.4}$$

The beat wavelength is equal to the distance that the two component waves, originally in phase, have to travel to get out of phase and then back into phase again. Figure 8.4 illustrates this definition graphically.

In order for the Fourier-transformed interferogram to resolve the two frequencies in the spectrum, the interferogram should contain at least *one*

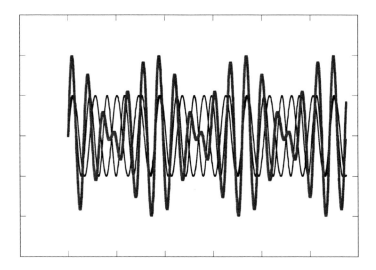

Figure 8.4 The sum of two sine waves (the lighter lines) of different frequency gives a characteristic waveform (the darker line) that has a *beat*. The wavelength of the composite beat wave is related to the difference in the frequencies of the component waves. The beat wavelength also dictates instrumental parameters necessary to resolve the two component signals after a Fourier transform.

complete beat. This means that the moving mirror should move by at least 1 λ_{beat}. This ultimately gives us our resolution criterion: the optical path difference should be equal to the beat wavelength.

$$\delta = \lambda_{beat} . \tag{8.5}$$

In terms of the difference in the frequencies of the two original signals, we have

$$\delta = \frac{1}{\Delta v} . \tag{8.6}$$

This is how we determine the relationship between resolution and instrumental factors: in order to resolve signals that differ by Δv (in cm^{-1}), the difference in the split light beams' pathlengths should be $1/\Delta v$ (in cm). Since the light in the moving-mirror path covers the distance between mirror and beamsplitter twice, the mirror's movement must only be half of δ. It is easy to see from this relationship that for signals that are 1 cm^{-1} apart, the mirror must move 0.5 cm; to resolve signals that are 0.50 cm^{-1} apart, the mirror must move 1 cm; and so forth. In practice, mirror movements are usually less than this, but it still represents a good rule-to-thumb for resolution on FT instruments.

The relationship between resolution and mirror movement explains why higher resolution becomes increasingly challenging. A higher-resolution spectrum requires that a mirror move farther but with higher accuracy, especially when averaging multiple scans to reduce noise. Higher resolution thus requires better optics, better stability, more care—and usually a higher price. Most commercial instruments can achieve resolutions of 1.0, 0.5, or even 0.25 cm^{-1} easily.

8.4 Noise: Sources

In spectroscopy, as with any measuring technique, the concept of noise has an important impact on what is actually been measured. In almost all spectroscopic techniques, some sort of signal—an intensity of a light beam, for example—is measured by a detector. In almost all measurements of this type, there are variations in the signal that are *not* due to the absorption/emission process. Instead, these variations are caused by various factors inherent in the process of the measurement itself. That is, they are noise.

We differentiate noise from signal, which is the desired measurement and in almost all spectroscopic techniques is electrical in nature. Zero signal might be zero current, but it may also be some arbitrary baseline current. Deviations from this baseline represent a signal to be measured. Noise is a fluctuation in a signal that tends to obscure the actual signal. A simple example is a person trying to say "hello" while a 100-piece orchestra is playing Mozart at full volume. A spectroscopic example would be a flicker in the source of a single-beam colorimeter, causing a fluctuation in the intensity of the light measured by the detector. In either case, an unwanted signal is imposing itself on a desired measurement. That is noise.

Noise has several sources, which we will discuss here. At some level, noise in spectroscopy is inherent because of the very nature of light and atoms. Apart from noise caused by the nature of the universe, what are the types of noise in a typical spectroscopic measurement?

We first differentiate between *random* and *nonrandom noise*. Random noise is as its name suggests: unpredictable. A flicker in a source lamp may be one type of random noise. Nonrandom noise has some sort of temporal pattern to it. For example, someone pacing back and forth next to his sensitive FTIR optical bench might be causing vibrations that can interfere with the measurement of a vibrational spectrum (as the author did once as a graduate student).

Interference is a type of noise caused by the surroundings imposing their characteristics on the spectroscopic system. For example, in the United States, alternating current has a frequency of 60 cycles per second; it is not uncommon to find nonrandom noise at that frequency and/or its multiples. Europeans use 50 Hz AC current, and so should be wary of noise at that frequency and its multiples. Similarly, any electronic frequency generator (say, a 2-kHz generator used in a conductivity experiments in a physical chemistry lab) has the potential for imposing interference noise on a spectroscopic experiment.

Some interference noise can be nonrandom. If a large electrical device on the same electrical circuit is turned on during an experiment, an electrical fluctuation can impose some interference noise. The passing of an elevator, as another example, can affect the performance of instruments nearby. Infrared signals between computers and printers and remote mice might be interfering with your IR measurement, although that is probably a highly improbable scenario.

Random noise that we cannot escape is sometimes referred to as *white noise*. For this type of noise, the amount (technically, its power) is independent of frequency and appears as a certain baseline level at any frequency.

Some random noise is frequency-dependent, however. Because it takes less energy to make low-frequency noise, there is more of this sort of noise at low frequencies than at high frequencies. Because of this, this type of noise is called *1/f noise*, with *f* representing frequency.

Finally, noise could be due to unwanted signal. This type of noise is especially problematic for two reasons. First, it may be difficult to identify the source of the unwanted signal. For spectroscopists who build their own apparatus, this can be a major obstacle. Second, while the source may be identifiable, the elimination of the source may be problematic. Again, astronomers and their battles with streetlights exemplify this problem.

What are the sources of these types of noise? We have already mentioned a few. Electrical power (no matter what the source—a wall outlet, a power supply, a battery) brings with it a potential source of noise. Unwanted signal is another source.

Other sources of noise are more insidious. They either have their origin in poor experimental setup or unrecognized experimental hazards. An apparatus for doing visible spectroscopy that has bright, shiny compartments is virtually begging for random reflections to add to the noise. Similarly, infrared spectra from spectrometers located next to mechanical vacuum pumps might be contaminated with vibrational noise as well. Even more invisible are the electrical circuits that we use to get our power. In many sensitive systems, if certain components are simply plugged into the same circuit in a room, there may be enough feedback noise to affect a measurement. Recognition of this sometimes requires that new electrical circuits be laid before proper experiments can be performed. Seldom does spectroscopy concern itself with building codes! But the nature and source of noise sometimes require it.

8.5 Noise: Minimizing

Minimizing noise in spectroscopic measurements is of varying concern to experimentalists. If one uses a commercial spectrometer, its very design should take noise-minimization factors into account (with a major exception). On the other hand, one who works with a homemade spectrometer will have to work to see that noise is minimized in either the design of the instrument or the performance of the experiment (or both). Let us discuss some ways in which spectroscopists minimize noise in a spectrum, with the understanding that different approaches are necessary for different types of spectroscopy, and that there will always be some noise in any spectrum.

Spectroscopists consider the signal-to-noise ratio (S/N) one indication of the quality of a spectrum. It is defined as the average value of the signal ratioed to its standard deviation. Most spectroscopic instruments need a signal-to-noise ratio of at least three (and preferably higher, of course) to support the identification of a spectral transition.

Isolate your system. As mentioned in the previous section, noise can be caused by electrical circuits, mechanical vibrations, or the presence of unwanted signals from other light sources or unwanted reflections (called *stray light*). If you isolate your spectrometer or light source or detector from these things, you can minimize their negative impact on a spectrum. For example, you can plug an electronic device into a dedicated electrical circuit to minimize interference from other electrical devices, or plug *all* devices into the same grounded circuit. Metal surfaces in the beam path of a spectrometer can be painted black to minimize stray reflections. Black cloth, opaque shields, and other materials can be used to block light from unwanted sources. Many companies sell massive tables that minimize transmission of mechanical vibrations; also, vibration-damping systems can be built to keep mechanical vibrations from affecting a measurement. (For example, a vibration-damping system was necessary for the developments of the first working scanning tunneling microscope.)

Increase source intensity. The more intense the light source is, the less the spectrum should be affected by unwanted light sources. For example, the use of a diamond anvil cell (DAC) in an infrared spectrometer blocks more than 99% of the infrared beam. Because of the low beam intensity reaching the detector, resulting spectra are very noisy. The situation can be improved by using special condensing optics designed to focus the IR beam to a smaller spot so that more light can pass through the DAC. Many spectroscopic accessory manufacturers sell optical assemblies for the specific purpose of improving light throughput and, therefore, increasing the S/N.

Spectroscopists performing attenuated total reflection (ATR) spectroscopy also have this problem and can get optics designed to increase the throughput of the light through the ATR element. The importance of source intensity is also illustrated by the changes in Raman spectroscopy with the development of the laser. Before the laser, Raman spectra were recorded using mercury or xenon lamps as the light source, and were measured over a time span of hours or days. Over such a time, sources of noise could have a larger effect on the final spectrum. Using lasers, so many more photons were available from the source that a Raman spectrum could be measured in much less time and with correspondingly less noise.

On a similar note, light sources that have a flicker also contribute noise to spectrum. The obvious solution is to minimize the flicker by using a more stable power supply or modifying the source itself so that it outputs a more stable light intensity.

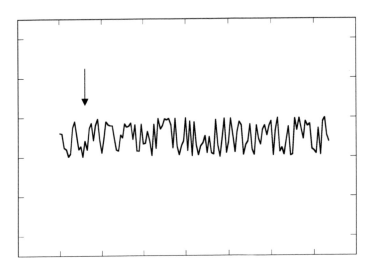

Figure 8.5 Random noise hides a spectral signal, which is not apparent in this simulated spectrum.

Decrease the spectral resolution. Though not the best solution, this is sometimes a necessary trade-off if no other option is available.

Average multiple scans. This is the major exception that was mentioned at the beginning of this section. Analog recording of a spectrum (that is, strip charts) is decreasing in popularity, because of the prevalence of personal computers. Because of this, multiple spectral scans of a sample can be measured, stored in computer memory, and then averaged. The averaging decreases the amount of random noise, which tends to cancel itself out if it is truly random. A well-known rule is that the S/N increases proportionally to the square root of the number of scans. Thus, a spectrum that results from 16 scans should have a S/N that is four times higher than a spectrum from a single scan. The spectroscopist usually has the ability to specify how many scans are measured and averaged. Spectrum averaging is especially common in FTIR spectroscopy.

Figures 8.5 and 8.6 show the effect of spectrum averaging. Figure 8.5 shows a generated spectrum with the signal (at the position marked by an arrow) that is effectively lost in the noise. However, when 32 of the spectra are measured and averaged, the level of noise is decreased and the signal is more apparent. Although Figs. 8.5 and 8.6 are contrived examples, they are exact parallels of what occurs when measuring spectra.

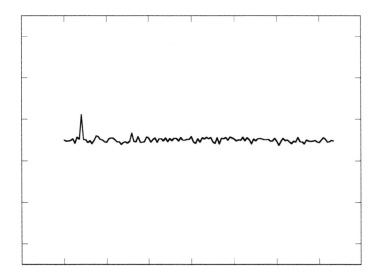

Figure 8.6 When 32 simulated spectra are averaged, the randomness of the noise cancels a lot of it, allowing the signal to be visible. This figure has the same scale of Figure 8.5. More scans would reduce the random noise even more, although at the expense of more time and diminishing returns.

There is a trade-off with multiple scans, too. It takes more time to collect more spectra. In addition, because the increase in the signal-to-noise ratio is proportional to the square root of the number of scans, the improvement in the spectrum slows as the number of scans increases. To improve the signal-to-noise ratio of Figure 8.6 by a factor of 2, we will need to measure 32 × 4 (128) scans and average them. However to improve the S/N by a factor of 10, 32 × 100 (3200) scans will have to be measured. Realistically, there will always be a balance between effort and noise.

References

1. STM Publication E131 – 94, "Standard Terminology Relating to Molecular Spectroscopy", 1994.
2. P. R. Griffiths, J. A. de Haseth, *Fourier Transform Infrared Spectrometry*, J. Wiley & Sons, New York, 1986.
3. J.D. Ingle, Jr., and S.R. Crouch, *Spectrochemical Analysis*, Prentice-Hall, Englewood Cliffs, NJ, 1988.
4. D. A. Skoog, D. M. West, F. J. Holler, *Analytical Chemistry: An Introduction*, Saunders College Publishing, Philadelphia, 1994.
5. B. C. Smith, *Fundamentals of Fourier Transform Infrared Spectroscopy*, CRC Press, Boca Raton, FL, 1996.

INDEX

David W. Ball has been an Associate Professor of Chemistry at Cleveland State University since 1990. He received his B.S. in chemistry from Baylor University and a Ph.D. from Rice University. Prior to moving to the Cleveland area, he conducted post-doctoral research at Rice and Lawrence Berkeley Laboratory. To date, he has almost 100 publications, equally split between research papers and articles of educational interest. His research interests include low-temperature infrared spectroscopy and computational chemistry of weak molecular complexes and high-energy materials. He has edited or written five books, including a math review book for general chemistry students and *The Basics of Spectroscopy*. He has been a Contributing Editor to *Spectroscopy* magazine since 1994, writing "The Baseline" column. He is active in the American Chemical Society, having served as Cleveland Section chairman and a member of the national ACS Committee on Chemistry and Public Affairs. His other interests include wine and beekeeping. Dr. Ball is married and has two sons.